Mapping Is Elementary, My Dear

Mapping Is Elementary, My Dear

100 Activities for Teaching Map Skills to K–6 Students

S. Kay Gandy

ROWMAN & LITTLEFIELD
Lanham • Boulder • New York • London

Published by Rowman & Littlefield
An imprint of The Rowman & Littlefield Publishing Group, Inc.
4501 Forbes Boulevard, Suite 200, Lanham, Maryland 20706
www.rowman.com

6 Tinworth Street, London, SE11 5AL, United Kingdom

British Library Cataloguing in Publication Information Available

Library of Congress Cataloging-in-Publication Data

Names: Gandy, S. Kay, 1954– author.
Title: Mapping is elementary, my dear : 100 activities for teaching map skills
    to K–6 students / S. Kay Gandy.
Description: Lanham, Maryland : Rowman & Littlefield, 2020. | Includes
    bibliographical references. | Summary: "The book includes a suggested
    scope and sequence for teaching map skills in the elementary grades and a
    glossary of geographic terms"—Provided by publisher.
Identifiers: LCCN 2020006350 (print) | LCCN 2020006351 (ebook) | ISBN
    9781475856774 (cloth) | ISBN 9781475856781 (paperback) | ISBN
    9781475856798 (epub)
Subjects: LCSH: Maps—Study and teaching (Elementary)—Activity
    programs. | Map reading—Study and teaching (Elementary)—Activity
    programs.
Classification: LCC GA130 .G36 2020 (print) | LCC GA130 (ebook) |
    DDC 372.89/1—dc23
LC record available at https://lccn.loc.gov/2020006350
LC ebook record available at https://lccn.loc.gov/2020006351

Contents

A chance encounter with a trash can changed how I teach. One day at the end of a school year, I happened to walk past the principal's office and spotted a thick book in the trash can entitled *Summer Workshops for Teachers*. I pulled it out, scanned its pages, and found an offering for a three-week summer workshop sponsored by the National Geographic Society. I loved reading *National Geographic* magazines and sharing what I learned with my students. So, I applied and was accepted to the workshop. I had no idea this would completely change my life and goals.

Those of us who attended the training became "teacher consultants," but in my words, we became "geo-evangelists." We were trained in geography content and how to conduct professional development for teachers. The National Geographic Society planned a grassroots movement to get geography back into the classroom. All fifty states (along with Puerto Rico and Canada) created geographic education alliances with trained teacher consultants. Each teacher consultant was tasked with developing exciting geography lessons to share with others. Many of the activities in this book are taken from lesson ideas shared by teacher consultants I encountered in workshops across the nation.

I did further training with the National Geographic Society and presented workshops as far away as New Zealand. When I retired as a classroom teacher, my work with professional development inspired the completion of my doctorate degree, thus securing a position at a university training future teachers. I later earned a master's degree in geography and developed a geography course for teachers. For several semesters, I worked with the gifted program at the university to offer Map Mania courses on the weekends for K–6 gifted students. I also served as co-coordinator for the state geographic alliance. All these initiatives inspired lessons that make up this book's content.

My accidental discovery in the trash enlivened my classroom and changed the course of my career. This book is a love letter to all the state geographic alliances and teacher consultants who have affected hundreds of thousands of students over the past two decades. Just remember . . . without geography you are nowhere!

Why Geography?

Most states include geography standards within the social studies strands, yet geography is sometimes not perceived as an essential element in the curriculum. Through geography instruction, teachers can show that maps are useful tools to help the young reader put stories into perspective, the young mathematician read gridlines and understand scale, the young citizen make wise decisions about the environment, and the young scientist understand climate change. Geography can tell you where you are and how you got there, how the earth affects you, and how you affect the earth.[1]

Geographers look at the world in spatial terms. They are interested in where things are located on the earth and why. Maps are tools that geographers use to communicate information about places. Physical geographers are interested in the study of natural features on the earth's surface, which connects with earth science. Human geographers are interested in man-made features, where people live, and why they choose to live there. Geography brings an understanding of ourselves, our relationship to the world, and our interdependence with others.

There are many reasons that schools may not emphasize the teaching of geography. Often, teachers are underprepared to teach geography to students. Many elementary teacher-training programs do not require geography courses, or instead offer teacher candidates the option to choose between geography and history. Testing mandates override concerns of teaching geography in some grade levels. Poor instructional practices also contribute to lagging geographic knowledge. For example, teachers may expose students to one map projection with distorted land sizes and never teach the discrepancies between the representations of a globe to a flat map. The forty-eight contiguous states may be presented as one scale on a map, while Alaska and Hawaii are designated with different scales. It can be quite confusing when teachers state that Alaska is bigger than Texas, when students can see on the map it is not. In addition, the definition of geography is often misunderstood. Many teachers consider geography to be only the use of maps, or they classify geographic concepts as earth science. Even if geography is taught in elementary schools, students may not understand they are learning about geography, therefore their concept of geography is limited to place names, locations, and physical landforms.[2]

In textbooks, relationships to real places are seldom explained. Many geographic activities just involve the coloring of maps or finding a location on a map. Activities such as these have no real meaning and do not help students understand mapping skills. Students should have the chance to explore and understand where they live and all the places surrounding them to make sense of their world. They should develop knowledge and understanding of physical elements of geography, such as weather, climate, and landforms; human elements of geography, such as populations and settlement; and environmental elements of geography, such as the use and misuse of natural resources.[3]

Through geography, students can feel a connection with people they have never met and places they have never been. They can understand how people and places came to be and how they affect one another. Teachers can encourage students to identify reasons for the physical and human characteristics of place, to question the accuracies of maps, and to make connections between places. Geography can inspire children to care about and commit to their place and community.[4]

A good place to begin is to introduce geographic skills to students: asking geographic questions, acquiring geographic information, organizing geographic information, analyzing geographic information, and answering geographic questions. These skills translate well to other subject areas. Teachers can begin with an essential question, such as, "Where is the best place to build a new school in our community?" Have students acquire information about the locations of other schools, population statistics, land availability, transportation patterns, and so on. Students would then organize the information into charts and graphs, analyze for patterns, and make connections. Finally, students would answer the questions and perhaps present their results to a public entity.

In the early grades, students should distinguish the difference between land and water on a globe, use a compass to find directions, understand symbols and legends on a map, investigate natural resources and physical regions, and be able to create a map of a familiar place. Geographic skill instruction should include asking students to determine directions, such as up, down, over, under, and beside. It should also include asking students to compare distances, such as long, short, near, and far. Students move on to learn to use special purpose maps to identify and gather data, identify the geographic characteristics that distinguish regions, and use scale to determine distances.

Teachers need to learn how to go beyond having students memorize states and capitals and obtain resources and materials that recognize the importance of geographic education. Students will use critical thinking skills when they question the accuracy of maps, identify biases of the cartographer's choice about what to include or omit on maps, consider the continents or countries placed in the center of maps, examine distortions of different map perspectives, and note the orientation of maps.[5] Handheld devices, such as phones, are familiar to students as navigation tools, but what tools did navigators use hundreds of years ago? Celestial navigation, the astrolabe, cross staff, sextant, compass, and other geographic tools can be interesting to students in their search for geographic understanding.

The goal of this book is to provide basic information about geography through mapping activities for elementary students to teach specific geographic skills. Each chapter contains basic information about the concepts, suggestions for activities to teach the concepts, and children's literature that relates to the concepts. The chapters conclude with assessment questions that teachers can use to ensure that students understand the concepts.

Chapter 1 will focus on location to expose students to activities to find precise and comparative points of location. Chapter 2 will introduce students to perspectives and how to look at the world from different viewpoints. Chapter 3 will spotlight scale to provide information about representing real places on a small piece of paper. Chapter 4 will provide students with orienteering information. Chapter 5 introduces abstract thinking with map symbols and map keys. Chapter 6 will familiarize students with a variety of types of maps and their uses. Chapter 7 will relay ways to use technology with geography, most particularly through GIS and GPS. Chapter 8 will introduce mapping skills with the Five Themes of Geography (Location, Place, Human/Environment Interaction, Movement, and Regions). The activities are written in short paragraphs to give the busy teacher an opportunity to scan for quick lesson ideas. The book ends with a suggested scope and sequence for teaching geography in appendix A, a glossary of map terms in appendix B, graphs and charts for activities in appendixes C through I, and a summary list of all activities within the chapters.

Teachers are the guides to encourage students to question the accuracy of maps, to examine places from different perspectives, to trace the routes traveled by literary characters, to understand why places are located where they are, or to compute distances between places. Providing concrete experiences involving real environments will provide a firm foundation for improving students' cognitive spatial abilities.[6] Teachers need to show students that geography is everywhere.

NOTES

1. Joe Rhatigan and Heather Smith, *Geography Crafts for Kids: 50 Cool Projects & Activities for Exploring the World* (New York: Scholastic, Inc., 2002), 10.

2. Mary Haas, "Teaching Geography in the Elementary School," *ERIC Digest*, ED309133 (1989): 2–7.

3. Lyn Malone, Anita M. Palmer, and Christine L. Voigt, *Mapping Our World: GIS Lessons for Educators* (Redlands, CA: ESRI Press, 2002), 103.

4. David Sobel, *Mapmaking with Children: Sense of Place Education for the Elementary Years* (Portsmouth, NH: Heinemann, 1998), 9.

5. Ava L. McCall, "Promoting Critical Thinking and Inquiry through Maps in Elementary Classrooms," *The Social Studies* 102, no. 3 (2011): 132–38.

6. Sharon Pray Muir, "Understanding and Improving Students' Map Reading Skills," *Elementary School Journal* 86, no. 2 (1985): 207–16.

## CHAPTER 1

Location

Figure 1.1. *Absolute and Relative Location*
Madalyn Stack, artist

The activities in this chapter explore ways to teach the concept of location. Location tells us where a point is on the earth and why it is there. Students should know that maps are tools they can use to find out where things are located on the earth. Every spot on the earth has a unique location. Begin instruction with students by exploring synonyms for location: site, locale, spot, situation, whereabouts, point, setting, environment, area, place, address, bearings, and venue. Encourage students to start where they are and learn their own place before venturing out into the world.

When teaching about location, ask students where something is (city, state, country), then talk about what it is like there (e.g., hot, cold, rural). Use maps and globes to find the locations. You might show photographs of the places from different perspectives. Start with places the students know, such as their home, neighborhood, and classroom. Typical concepts and skills required in instruction for location include locating states, countries, continents, and oceans on a map; locating main cities and capitals; locating regions; locating the prime meridian, international date line, and time zones around the world; locating explorers' routes, colonies, and resources; and locating natural and cultural features of places studied.

Two terms students need to know are "absolute" and "relative" location. Every feature on Earth is located at a precise or absolute point. Absolute location can be determined by using the mathematical grid system of latitude and longitude. Students could practice identifying the absolute location of places: Our classroom is number 204 at Summerfield Elementary, on 321 South Main Street in Monroe, Louisiana. New Orleans is located at 30 degrees north, 90 degrees west. The location of one place compared to another is known as relative location. You could have students practice describing places with relative location: Our classroom is near the water fountain, across from the library, or last in the hallway.

Mapmakers often use a set of imaginary lines, called a grid, to divide space on maps. Horizontal lines have letters, and vertical lines have numbers. This allows the reader to find the location of any place on the map by examining where the horizontal and vertical lines meet (e.g., B7 or D8). The grid system on Earth is made up of two sets of imaginary lines. Parallels of latitude run east and west around the globe, while meridians of longitude run north and south. Latitude and longitude are measured in terms of the 360 degrees of a circle and subdivided into minutes and seconds. So, a measurement of 42°53'52" N would be read as 42 degrees, 53 minutes, 52 seconds north of the equator.

The prime meridian is 0° longitude and passes through the site of the Royal Naval Observatory in Greenwich, England. Longitude lines east of the prime meridian are numbered 1 through 179 (eastern hemisphere). Longitude lines west of the prime meridian are numbered the same (western hemisphere). The 180 degree line is called the international date line and is exactly halfway around the earth from the prime meridian.

The equator is 0° latitude. Latitude lines are numbered from 0 to 90 from the equator to the North Pole (northern hemisphere) and from 0 to 90 from the equator to the South Pole (southern hemisphere). Latitude lines are parallel and never meet. Sailors measured latitude by length of day, height of sun, or known guide stars; they measured longitude by each hour's time difference between the ship and homeport (15 degrees). Equipment used to measure latitude and longitude included the mariner's astrolabe (marked positions in the sky to determine latitude); cross staff (measured altitude of stars, which allowed sailors to find latitude, tell time, and find direction at night), and chronometer (determined longitude at sea).

Activities to learn location can be as simple as playing with toy cars on carpet roads, pushing toy trucks in the sandbox, laying out the neighborhood with milk carton houses, or building Lego villages. Playing Bingo can introduce skills needed to read a grid, while dancing the "Hokey Pokey" provides practice with directional terms. Kinesthetic activities lay the foundation for spatial learning.

ACTIVITIES FOR TEACHING LOCATION

1. Toss the Globe

Materials: inflatable globe, small sticky notes to mark land (green) or water (blue) on a graph
Time: 15 minutes
Suggested Grade Levels: K–2; 3–6

Have the younger students sit in a circle and toss an inflatable globe of the world to a student to catch with both hands. Ask the student, "Is your right thumb on land or water?" Using sticky notes (blue for water, green for land), let the children make a pictorial graph of the results. It will soon become obvious to students that the earth has more water than land. Students should note that the right thumb tends to land on water more often than land, as about 71 percent of the earth's surface is water.

Move to more challenging questions with older students. In what hemisphere(s) is your right thumb? Which direction would you travel from your thumb to reach South Dakota? What language is spoken in the country nearest your right thumb? What currency is used in the country nearest your right thumb?

2. Your Place in the World

Materials: pre-cut concentric circles or materials to make a flip book, book *Me on the Map* (Sweeney, 2018), book *Mapping Penny's World* (Leedy, 2003), drawing paper, drawing tools
Time: 30 minutes
Suggested Grade Levels: 1–2

Read the *Me on the Map* book to the students. This story indicates how a young girl recognizes her place and space in a bedroom, home, street, neighborhood, city, state, continent, country, and world. (It begins with the place the young learner is most familiar with and branches out to the world.) Ask the students these questions: "What city do you live in? What state do you live in? Which states border your state? What country do you live in? What continent do you live in? What hemisphere(s) do you live in? Which planet do you live on?" Pre-cut concentric circles (or make a flip book) and note locations for My Street, My City, My State, My Country, My Continent, and My Planet. Allow the students to draw and label each place. Display completed products on a bulletin board. An alternate activity can be to have students use a series of printed maps and stickers to indicate their locations with school, neighborhood, town, county, state, country, continent, and globe. A similar story is *Where Do I Live?*, which takes children from their bedrooms to the universe and back.

Another book to map the familiar and introduce map symbols is *Mapping Penny's World*. Lisa learns the parts of a map at school and is given the assignment to create maps at home. With the help of her dog Penny, Lisa maps hideouts, routes to Penny's friends, hiking trails in the park, favorite places in the neighborhood, and Penny's imaginary trip around the world. After reading the book, have students draw a map of the school and locate important places, such as the cafeteria, the library, or the principal's office. Encourage students to make their own home and neighborhood maps.

3. Where Are You Now?[1]

Materials: 4 meter sticks, 2 skeins of red yarn, 2 skeins of black yarn, 4 rolls of masking tape, 2 sets of index cards with letters A–J (one letter on each card), 2 sets of index cards with numbers 1–10 (one number on each card), paper for each student
Time: 45 minutes
Suggested Grade Levels: 1–2; 3–5

Ask students to describe their location in the classroom (e.g., near the front, in the middle, by the door). Students must use different location words than the person(s) who spoke before them. Discuss relative location (position relative to a landmark). Teach positional words: left, right, above, below, beside. Next, ask students, "What is the absolute location (fixed position) of the classroom (e.g., room 101)? of this building? of this school? of this city?" Discuss that latitude is measured north or south from the equator from 0 to 90 degrees and longitude is measured east or west from the prime meridian 0 to 180 degrees. Inform students that every home has a street address and every place has a global address, which can be identified by latitude and longitude.

Divide the class into two teams to "grid" the classroom. Use meter sticks, yarn, tape, and cards with numbers (1 to 10) and letters (A to J). One team of students would tape the number cards a meter apart across the front and back walls of the room in the longitude positions (1 to 10 on one wall, 1 to 10 on the other wall mirroring it). The other team would tape the letter cards down the sidewalls of the room for the latitude positions. Connect the cards by yarn to form a grid (e.g., A to A, B to B; 1 to 1, 2 to 2). Grid the yarn over the heads of students. When each team has completed the grid system, have each student draw a grid on paper and determine his or her absolute location in the classroom. Ask students again to describe their location in the classroom, this time using latitude and longitude measures. Be sure to state the latitude first.

You might choose to do this activity outside with younger students. Students can help create a grid with pegs and string on the playground, and then place themselves at various positions according to the teacher's statements ("Sam, place yourself at E4," or "Where is Callie located on the grid?"). Younger students might grid the classroom on the floor level. Often, teachers use the one-foot square tiles on the floor as the measurement tool to grid the classroom.

4. Where Are You Going? Where Have You Been?

Materials: drawing paper, pencils
Time: 25 minutes
Suggested Grade Levels: 1–5

Have students map routes going to the cafeteria at school, practicing a fire drill at school or home, or going home after school. Students should note cardinal directions and locate landmarks. You might take a walk around the school neighborhood and have students reconstruct on a map what they saw and heard on the walk. Ask the students, "Why do you think the school is located at this place? If we were to build a new school, where would be a good location?"

5. Travel Stories

Materials: drawing paper and pencils; books of travels, such as *The Journey of Oliver K. Woodman* (Pattison and Cepeda, 2009), *Three Days on a River in a Red Canoe* (Williams, 1984), *Paddle-to-the-Sea* (Holling, 2004), or *The Incredible Journey* (Burnford and Burger, 2018)
Time: 45 minutes
Suggested Grade Levels: 4–6

Read stories that tell of journeys across land and sea, and then have students draw maps of the journeys. A good example is *The Journey of Oliver K. Woodman,* which tells the story of an uncle unable to get away for a visit to his niece, so instead he builds and sends a wooden man in his place. A note in Oliver's pouch asks people to help him get to his destination. *The Incredible Journey* tells the story of three house pets who cross the country and make their way home to the family they love. In the book *Three Days on a River in a Red Canoe*, students can follow the adventures of Mother, Aunt Rosie, and two children as they take their canoe down the river. A Caldecott Honor Book, *Paddle-to-the-Sea* describes the story of a Native American boy who carves a wooden canoe and writes an inscription that the canoe is trying to find the quickest route to the sea. Maps of the journey are included as the canoe spends four years before finally making it to the sea. Review parts of a map that should be on the students' drawings: key (or legend), title, scale, compass rose.

6. Music across America[2]

Materials: blank map of the U.S. with state outlines, snippets of songs that mention places (e.g., "Take Me Back to Chicago," "California Girls," "My Old Kentucky Home," "The Night the Lights Went Out in Georgia," "The Yellow Rose of Texas," "Louisiana Saturday Night").
Time: 20 minutes
Suggested Grade Levels: 4–6

Give each student a blank map of the U.S. with state outlines. Play snippets of songs that mention places and have the students mark the places on the map. Have students brainstorm other songs that mention location. Then have students categorize songs that mention cities, states, or regions. Many songs mention a variety of places within one song (e.g., "This Land Is Your Land" and "God Bless the USA"). Challenge students to find songs that mention a place from each of the fifty states.

7. Where Do Products Come From?

Materials: book *How to Make an Apple Pie and See the World* (Priceman, 1996), recipes, apple pie, Snicker's candy bars, world map for each student; different types of breads, book *Bread, Bread, Bread* (Morris and Heyman, 1993)
Time: 25 minutes for each lesson
Suggested Grade Levels: 3–6

Read the *How to Make an Apple Pie and See the World* book to students. When the market is closed, a young girl travels around the world to collect the ingredients to make an apple pie. She takes a steamboat to Italy to gather wheat, a train to France to get an egg, a boat to Sri Lanka to get cinnamon, a ride to England for fresh milk from a cow, a banana boat to Jamaica to cut sugar cane, and a parachute to Vermont to pick apples. Give each student a world map to trace the journey as you read. Make picture cards of forms of transportation and ingredients for the pie and have students place them on a map of the world.

Have students bring in their favorite recipes and determine where in the world the ingredients might come from. Following the format of the book, students could create their own books and maps of how to make a particular food. A recipe for apple pie is included with the book, so students can have a tasty treat for the end of the lesson. The author has also published a book called *How to Make a Cherry Pie and See the U.S.A.* With various modes of transportation, the young girl travels throughout the U.S. in search of materials to make baking tools.

Choose a product, such as a Snicker's candy bar, and have students trace the journey of where each ingredient comes from (e.g., peanuts from the Sudan, corn syrup from Iowa, sugar from Hawaii, chocolate from Guyana, vanilla from Mexico) on a world map. Note that the Snicker's bar factory is in New Jersey. Have the students pretend that they are the CEO of the Snicker's factory. Discuss with students what might happen if a tragedy happened in a country and one of the ingredients was no longer available (e.g., civil war in the Sudan, hurricane in Hawaii, severe drought in Iowa). Have students trace the journey of the Snicker's bar to their hometown.[3] NOTE: For students with peanut allergies, choose a different candy bar.

Have students get a partner and examine each other's clothing labels to determine what countries made the products. Ask students, "Which countries appear the most on the clothing labels? Why?" Trace routes on a world map of how products from other countries get to the U.S. Bring in a variety of tank tops from a single store, such as Old Navy. Have students look at the labels. Ask students to explain why the shirts are made in different countries. Talk about interdependence among countries.

The book *Bread, Bread, Bread* introduces breads from other cultures (e.g., tortillas, baguettes, chapattis) in sixteen different places around the world. A mapping activity can lead to discussions of how the breads get to markets in the United States or other countries. Students could locate wheat-growing areas on a map. Be sure to have different types of breads for students to taste!

8. Family News Night[4]

Materials: world map, news clips
Time: 20 minutes
Suggested Grade Levels: 4–6

Even the local news can be used as a teaching tool for mapping. This family geography challenge encourages parents, once a week, to find places on maps that are mentioned in the news and discuss characteristics of those regions with their children. Parents are encouraged to ask children questions such as, "Where is this city in the news located? What is it like to live in this place? How has the environment affected the way humans live in this place? How have humans affected the environment? What is changing in this place, and how is it changing?" Keep a world map in the family room to point out places as you discuss them.

Teachers can continue this activity in the classroom through shows such as PBS News Quiz,[5] which hosts headlines for the week and a series of questions to test student understanding, or CNN 10[6] for on-demand digital news.

9. Route Map[7]

> **Materials:** large pieces of art paper, construction paper, markers
> **Time:** 30 minutes
> **Suggested Grade Levels:** 3–6

Write a list of different stages of a journey you have been on recently or imaginary journey you would like to take. Make note of interesting landmarks you passed and the time as you passed. Draw a road going down the middle of a large piece of paper. Add clocks to show start and finish time of journey. Starting at the bottom of the paper, add construction paper symbols for different stages of your journey. Encourage the students to create their own route maps.

10. MIMAL—The Man in the Middle

> **Materials:** map of the U.S.
> **Time:** 15 minutes
> **Suggested Grade Levels:** 3–5

Tell the story of MIMAL (Minnesota, Iowa, Missouri, Arkansas, Louisiana)—the man in the middle of the United States—to teach students locations of states on a map. Point to each state as you tell the story. After the story, point to the states again and have students name the states. Challenge students to complete the story of MIMAL to include all fifty states.

MIMAL wears a hat because his head is in chilly *Minnesota*. You can see his face and nose in *Iowa*. MIMAL's pants are represented by *Arkansas*. He wears boots because his feet are in the Gulf by *Louisiana*. MIMAL must loosen his belt because he ate too much and is in misery

(*Missouri*). MIMAL faces the Great Lakes so he can watch over lovely HOMES (Huron, Ontario, Michigan, Erie, and Superior). MIMAL turns around and climbs on his *Texas* saddle to visit the four corners because he can and you can—U CAN (*Utah, Colorado, Arizona, New Mexico*). But MIMAL does not ride a horse. He rides a COW (*California, Oregon, and Washington*). MIMAL is cooking up some *Kentucky* fried chicken in his *Tennessee* skillet.

11. We Are the World

Materials: book *Chester the Worldly Pig* (Peet, 1978), large pieces of art paper cut into shapes of continents, atlases for each student

Time: 40 minutes

Suggested Grade Levels: 2–4

Read the book *Chester the Worldly Pig* to students. Discuss how Chester gets to join the circus in an unexpected way through the unusual shape of his spots, which display the continents of the world. Divide students into seven groups and assign each group a continent to research in atlases. Have the students color and draw physical features and animals on large pieces of paper cut out in the shapes of the seven continents. Then place the continents on the wall in the correct location to form a flat map.

ANNOTATED CHILDREN'S LITERATURE FOR TEACHING LOCATION

Burnford, Sheila, and Carl Burger. *The Incredible Journey*. New York: Yearling, 2018.
Three pets are separated from their owners and make an incredible journey back to their home. In their journey through the wild, the animals face wild animals, exposure to the elements of nature, and starvation.
Chesanow, Neil. *Where Do I Live?* Hauppauge, NY: B.E.S. Publishing, 1995.

Similar to *Me on the Map*, this book begins with a child's room and travels to the entire universe and back again. The explanations and details are more prevalent in this book, but the maps are not as realistic. A mini quiz is included at the end of the book.

Dillemuth, Julie. *Camilla, Cartographer*. Washington, D.C.: Magination Press, 2019.

Camilla is distressed when a snowstorm covers paths and landmarks and she can't use her maps to find her way. Camilla is inspired to make her own map as she clears pathways to the creek. The author gives information on spatial thinking and sketching and reading maps.

Elliot, David. *Henry's Map*. New York: Philomel Books, 2013.

Henry the pig is very concerned about the disorganization of the farm, so he creates a map to show every animal its proper place. When Henry leads the animals up a hill to compare the map to the farm, he finds that none of the animals are where they are supposed to be. The amusing book can inspire young cartographers to create their own maps.

Holling, Clancy. *Paddle-to-the-Sea*. Torrance, CA: Sandpiper Books, 2004.

A Caldecott Honor Book, this book describes the story of a Native American boy who carves a wooden canoe and writes an inscription that the canoe is trying to find the quickest route to the sea. He then sets the canoe on a snowbank near a river that eventually leads into Lake Michigan. The canoe spends four years on the water, and is picked up by loggers, anglers, and families before finally making it to the sea. Maps of each of the Great Lakes are included on most pages, and a map of the journey is included at the end of the story

Keller, Laurie. *The Scrambled States of America*. New York: Henry Holt and Company, 1998.

Kansas is bored and decides to throw a party for the other states with his friend Nebraska. At the party, the states come up with the idea to switch places. Once the excitement died down, the states realized that they weren't very happy with their new locations (Florida was freezing in the north and Minnesota got a sunburn), so they packed up and went back to their very own homes.

Leedy, Loreen. *Mapping Penny's World*. New York: Henry Holt and Company, 2003.

After learning the elements of a map at school, Lisa draws maps at home, starting with her bedroom. With the help of her dog Penny, Lisa includes all the things that people will need to read the map, such as a scale and a key to the symbols she has used. Lisa then has the idea to map Penny's world. She includes maps of Penny's hideouts, routes to Penny's friends, hiking trails in the park, favorite places in the neighborhood, and Penny's imaginary trip around the world.

Morris, Ann. *Bread, Bread, Bread.* New York: HarperCollins Publishers, 1993.
The photographs in this book display the many types of breads enjoyed by people all over the world. Readers are introduced to bagels, tortillas, and baguettes, as well as people eating, baking, and selling breads.

Pattison, Darcy, and Joe Cepeda. *The Journey of Oliver K. Woodman.* Boston: HMH Books for Young Readers, 2009.
An uncle cannot visit his young niece, so he builds a life-sized wooden man to send in his place. A note in the wooden man's pocket asked people to help him get to his destination. The story follows the cross-country trek with beautiful illustrations.

Peet, Bill. *Chester the Worldly Pig.* Boston: HMH Books for Young Readers, 1978.
A young pig wants desperately to join the circus and features his tricks anytime a circus train goes by the farm. It is not, however, his bag of tricks that gets him noticed. Chester's spots are in the shape of the world's continents. He gets his wish to be a star.

Priceman, Marjorie. *How to Make a Cherry Pie and See the U.S.A.* New York: Dragonfly Books, 2013.
Similar to the author's apple pie book, a young baker travels the United States in search of ingredients such as coal, cotton, clay, and granite to create baking tools. Travel modes include taxi, riverboat, airplane, and train.

Priceman, Marjorie. *How to Make an Apple Pie and See the World.* New York: Dragonfly Books, 1996.
When the market is closed, a young girl travels around the world to collect the ingredients to make an apple pie. Travel modes include a steamboat to Italy to gather semolina wheat, a boat to Sri Lanka to peel bark from the corundum tree to make cinnamon, a banana boat to Jamaica to cut sugar cane, and a parachute to Vermont to pick apples. Bonus pieces include a map of the world and a recipe for apple pie.

Sweeney, Joan. *Me on the Map.* New York: Dragonfly Books, 2018.
This book begins with the place the young learner is most familiar with and branches out to include the world. In this introduction to maps and geography, a young girl shows readers herself on a map of her room, the map of her house, the map of her street, a street map of the town, a state map (Kansas), all the way to her country on a map of the world. Then the process is reversed until the girl is back in her room.

Williams, Vera B. *Three Days on a River in a Red Canoe*. New York: Green-willow Books, 1984.

In this book, students can follow the red canoe down the river in the adventures of Mother, Aunt Rosie, and two children. This book includes several maps of the river trip and interesting tips for camping. The map at the beginning of the book presents an overall view of the river, while a later map features a close-up of a particular area.

ASSESSMENT QUESTIONS—LOCATION

- How are the equator and prime meridian alike? How are they different?
  Possible Answer: The equator separates the northern and southern hemispheres, while the prime meridian separates the western and eastern hemispheres. The equator is a latitude line, while the prime meridian is a longitude line.
- Where is your state located? What states are located north, south, east, and west of your state?
  Possible Answer: My state is Kentucky. It is located in the south-central region of the United States. Kentucky is bordered on the north by Illinois, Indiana, and Ohio; on the south by Tennessee; on the west by Missouri; and on the east by West Virginia and Virginia.
- What challenges do cartographers face?
  Possible Answer: A cartographer must figure out how to represent a three-dimensional object into two dimensions and maintain proportional relationships. They must decide how much information should be included on the maps and design the information so that it is useful to readers. Size, shape, or distance will be distorted on a flat map.
- Name the continents and oceans.
  Answer: North America, South America, Europe, Asia, Australia, Antarctica, Africa. Indian Ocean, Pacific Ocean, Atlantic Ocean, Artic Ocean, Southern Ocean (newest named ocean basin that surrounds Antarctica).
- What is an example of relative location for our classroom?
  Possible Answer: Our classroom is located near the water fountain, around the corner from the library, and next to the computer lab.

- What is the name of the place that the 0-degree longitude line crosses?
  Answer: Greenwich
- What is the latitude and longitude of our school?
  Possible Answer: My school is located at 36.9854° N, 86.4561° W.
- Your state wants to choose a new capital. Choose a city and discuss the effect of location on your choice as the new capital.
  Possible Answer: I think that Louisville should be the new capital of Kentucky because it is centrally located within the state. It has a thriving economy and could be easily reached by most of the populace.
- Find Africa on a map. How many latitude lines run through Africa?
  Answer: Five lines of latitude, including the equator.
- Use a map to identify the city you find at 30° N, 90° W.
  Answer: The city is New Orleans, Louisiana.

NOTES

1. Adapted from the lesson "Tic-Tac-Room" in *Finding Your Bearings* (Fresno, CA: Aims Educational Foundation, 1994), 103–5.

2. Adapted from lesson ideas by Carolyn Brugmann, Paula Nunez, and Carolyn Santangelo, teacher consultants with the Louisiana Geography Education Alliance (LaGEA).

3. Adapted from lesson "Snicker Fun" by Barbara Fields, teacher consultant with the Geographic Educators of Nebraska (GEON).

4. Adapted from lesson "Family Geography Night," developed by the Michigan Geographic Alliance, accessed November 4, 2019, https://www.cmich.edu/colleges/se/Geography/Michigan%20Geographic%20Alliance/Geography%20Outreach/Documents/FGNHandbook.pdf.

5. PBS News Quiz can be found at https://www.pbs.org/show/news-quiz/episodes/.

6. CNN 10 can be found at https://www.cnn.com/cnn10.

7. Adapted from lesson "Make a Route Map," Andrew Haslam, *Maps: Make It Work!* (Chicago: World Book Inc., 1996), 29.

## CHAPTER 2

Perspective

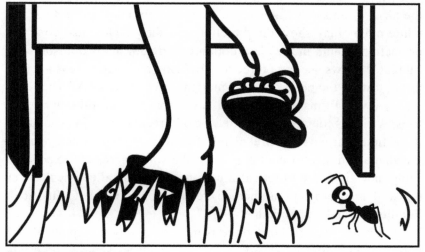

Figure 2.1. *Perspective*
Madalyn Stack, artist

The activities in this chapter explore ways to teach the concept of perspective. Since young children usually view the world from the ground level, it may be difficult for them to imagine an aerial view. Yet, maps are typically presented from this perspective. Begin by having students explore synonyms for perspective: outlook, view, standpoint, position, stance, slant, attitude, approach, and interpretation.

Spatial thinking is critical for students to develop as they learn geography. Elementary students typically learn concepts and skills related to perspective that include comparing a flat map to a globe and learning about mental mapping, sizes of land types, bird's-eye view, and types of map projections. Helping students build accurate mental images will lay the foundation for map reading and interpretation skills.

Maps are two-dimensional, yet the earth is three-dimensional. By placing a round earth on a flat map, cartographers face the challenge to make land areas meet. In order to deal with this problem of distortion, geographers use a variety of map projections. Map projections attempt to show the round earth as a flat model; however, distortions always occur in shape, size, or position. Map projections will be different based on the needs of the user. Some projections show the size of land and water areas as accurately as possible to scale, while others may show the direction and shape. The most common projections found in the classroom are Mercator and Robinson. The Mercator shows accurate direction, but land and water areas are distorted toward the poles. North America looks as big as Africa, yet in reality it will fit inside Africa and have room for several other countries. Africa is fourteen times the size of Greenland. China should be three times larger than Greenland. The Robinson accurately shows the shape and size of continents, but the water areas are expanded. Longitude lines are elliptical arcs rather than parallel lines. Other types of map projections include Peters, Conic, Polar, Cylindrical, Polyhedral, and Azimuthal.

Geographically speaking, there is no reason why a map should be north or south oriented. There is no "up" in space. Maps could easily show east or west as "up." Maps typically center on the prime meridian, or 0-degree longitude. Early mapmakers wanted Europe as the center of the world. Some maps placed Jerusalem at the center of the world. Placing Europe on the left side of a map is just as correct as placing it on the right.

Activities to learn perspective can be as simple as viewing different map projections, creating earth models in different shapes (triangular, hexagonal, donut), or pretending to be stuck on the ceiling. The important idea is to expose students to various perspectives and improve their spatial thinking skills.

ACTIVITIES FOR TEACHING PERSPECTIVE

12. Earth Models

Materials: globe, large paper grocery bags, newspaper, masking tape, blue tempera paint, paint brushes, construction paper, scissors, glue, yarn
Time: 45 minutes
Suggested Grade Levels: 2–4

Examine a globe with students. Name the continents and oceans. Students can make a model of the earth by stuffing large paper grocery bags with newspaper and taping with masking tape into a rounded shape. An alternative could be to use balloons and papier-mâché. Paint the model entirely with blue tempera paint and let it dry. Older students can cut out or tear paper shapes of the continents out of construction paper and glue them in the appropriate places onto the model. Younger students can use pre-cut continents. Tape yarn string to hang the globes from the ceiling.

13. Flat Map vs. Globe

Materials: orange for each student, safe orange cutter, paper towels, permanent markers
Time: 45 minutes
Suggested Grade Levels: 3–6

Give students an orange, an orange cutter, a paper towel (to dry off the orange), and a permanent marker. Have students draw the continents on the dried orange to represent the earth. Cut the orange around into four areas. Carefully peel the skin off and lay each piece flat. Ask students, "What is missing? (There are gaps in the land area of continents). What do cartographers do with flat maps to fix the gaps?

(Resize land areas) What is the difference between a globe and a flat map? Which is more accurate?" (Globe)

14. Food as Earth

Materials: grapefruit or orange; rubber bands; peach, boiled egg, or apple and cinnamon candies; fruit roll-ups; cookie map or popcorn balls; Neapolitan ice cream, straws; variety of puddings, cake mixes, or other types of foods that can be layered
Time: 45 minutes
Suggested Grade Levels: 2–4

Use a round fruit, such as a grapefruit or orange (or even a pumpkin). Place rubber bands around the fruit to show students the hemispheres, equator, and prime meridian. Use a peach or boiled egg to demonstrate the parts of the earth (core, mantle, crust). You can also slice a cross-section of an apple and melt cinnamon candies in the center to demonstrate the layers of the earth. Use fruit roll-ups and have students cut the shapes of the seven continents. The continents can be placed on a flat cookie map or on a round popcorn globe. Use a straw in Neapolitan ice cream to demonstrate how geologists get core samples from the earth.

Have students come up with their own ways to represent parts of the earth with various types of food. For example, use different flavors of pudding or cake mixes to represent the layers of the earth. Assign students to create a model of the earth using food. Of course, they get to eat the results!

15. Map Projections

Materials: various types of map projections (e.g., Cylindrical, Conic, Azimuthal, Mercator, Robinson, Sinusoidal, Fuller, McArthur)
Time: 30 minutes

Inform students of different types of map projections. Cylindrical projections have straight coordinate lines with horizontal parallels. An example would be the Mercator projection. Conic projections are designed to be able to wrap around a cone on top of a globe. An example would be the Albers conic. Azimuthal projections feature great circle arcs and are beneficial for finding direction from any point on Earth using the central point as reference. An example would be an Orthographic projection. Ask students, "Which map projection is found in our classroom?" Have students examine the longitude and latitude lines on each type of projection to determine what is different. Have students compare the land sizes and ocean sizes on each projection. Compare the size of Greenland to South America on each of the projection maps. Discuss with students which projection is a more accurate representation of the earth and why. Use the comparison chart in appendix C with students and have them research the various projections. Teachers can show websites that display maps of different perspectives. The Twisted Sifter website displays forty unusual maps, such as "Countries That Do Not Use the Metric System," "McDonald's across the World," and "Global Internet Usage Based on Time of Day."

16. View from the Top

Materials: books *As the Crow Flies* (Hartman, 1993) and *The Fox Went Out on a Chilly Night* (Spier, 1994), aerial photograph, transparency overlay of map, drawing paper
Time: 45 minutes
Suggested Grade Levels: 1–2

Two children's books are very helpful in teaching about perspective. *As the Crow Flies* uses words and pictures to describe different geographical areas from the perspectives of an eagle, a rabbit, a crow, a horse, a gull, and the moon. At the end of the book, all maps join together to present the big picture from the view of the moon. *The Fox Went Out on a Chilly Night* presents various panoramic views of a town and countryside as a fox goes out to find supper for his hungry children. The book is based on the old folk song of the same name. After reading the books, have students imagine they are stuck to the ceiling of a room in their house (or the classroom) and must draw a map of the location of all objects in the room from this viewpoint. It is often helpful to take an aerial photograph and impose a map on top to show students how a map can represent a picture. Teachers could use a transparent overlay to trace major features on a photograph and display the results as a "map."

17. Earth Shapes

Materials: pictures of the earth in different shapes (https://nrich.maths.org/1363); templates for earth shapes (http://www.3dgeography.co.uk/make-a-globe), glue, scissors, colors
Time: 45 minutes
Suggested Grade Levels: 4–6

Ask students, "What is the shape of the earth?" (Sphere) Because the earth is flattened at the poles, it can be called an oblate spheroid. Have students use the website to examine pictures of the earth in different shapes (e.g., donut, cube, pyramid). Ask students, "What might change if the earth were a different shape?" (Orbit, seasons, rotation, population centers, gravity, landforms) Have students use the templates at the website to make their own earth shapes.

18. Learning through Play

Materials: carpet with road map design, toy cars, sandbox, Legos, building blocks, dollhouse, small milk cartons
Time: 30 minutes
Suggested Grade Levels: K–2

Teachers can secure room carpets with road map designs and allow students to play with cars and trucks. Talk about the perspective of looking down at the scenery. Playing hide-and-seek allows students to explore the topography of the schoolyard. Have students pretend they are birds and view the playground from this perspective. Other perspective play activities include drawing roads in the sandbox, building designs with blocks, creating Lego villages, and moving furniture around in a dollhouse. Lunchroom milk cartons can be used to represent buildings in a town. Working with models helps young children develop mental images necessary for mental mapping skills.

19. Mental Mapping

Materials: drawing paper, yarn, drawing materials, atlases
Time: 30 minutes
Suggested Grade Levels: 4–6

Introduce students to the concept of mental mapping (a person's perception of the world). Mental maps tell us how much we remember of our surroundings. Give students thirty seconds to draw a representation of the earth. Draw circles to represent each continent. The circles should reflect the approximate size and location of the continents. Have students label oceans and continents. Compare each student's perception of the world.

Give each student a piece of yarn. Tell them that they are cartographers and they will attempt to shape continents, countries, or states with the yarn. Use an atlas to check for accuracy.

Have them form the shapes of Kentucky and Louisiana to practice first. Students might note that Kentucky looks like a chicken leg and Louisiana looks like a boot.

Have students draw a map of a place they each have been, such as the mall. Compare each map drawing. What is different? Why are they different? Note that students do not visit every store or place in an area, and their mental maps are often created by their interests.

20. How Big Is Africa?

Materials: various map projections of the world, Africa Map
   Poster found at http://www.bu.edu/africa/outreach/curriculum/
   curriculum-guide/
Time: 30 minutes
Suggested Grade Levels: 4–6

Use the map poster to show students a perspective of China, Europe, and the U.S. fitting inside the continent of Africa. Discuss the size of Africa as represented on several map projections and a globe. Ask students, "What surprised you most about this poster? Why is it important to know the size of a continent or country?" Lesson ideas are included on the website.

21. News Shapes the World[1]

Materials: newspapers, continent outlines, glue, blue construction paper, scissors
Time: 45 minutes
Suggested Grade Levels: 4–6

Divide students into groups of seven. Assign students in each group a specific article to read from a newspaper (e.g., environmental, international, local, political). Have students summarize the articles that they read. Students will then trace the outline of a continent on the news articles. Use one article for each continent. Document the information on the chart in appendix D. Construct a world map by gluing each newspaper continent to a large piece of blue construction paper. Label the continents and oceans. Have students discuss what geographic issues shaped their continents. Label the finished product "News Shapes the World."

22. Maps and Math

Materials: state map with outlined counties and highways
Time: 20 minutes
Suggested Grade Levels: 4–6

Have students examine a state map and ask the following questions: "Can you find a county that closely resembles the shape of a rectangle? What other shapes can you associate with counties in this state? How many highways are numbered by multiples of ten? How could you measure the distance around the county? Approximately, how long is the major highway through the county?" Allow students to suggest other ways to use maps and math.

23. Body Mapping

Materials: state maps for groups, drawing paper, drawing tools
Time: 25 minutes
Suggested Grade Levels: 2–4

The ruler is not the only way to measure the length of an object; students can use their bodies to develop a measuring system. A span is the distance measured by a human hand from the tip of the thumb to the tip of the little finger. A cubit is based on the length of the arm from the elbow to the tip of the middle finger (equivalent to two spans). A digit was considered the width of the middle fingertip. Have students measure the distance between two cities on a state map using a digit, a span, and a cubit measure. Compare measurements and note differences. Challenge the students to create a map using the scale of "digit." Ask students, "How does the scale change the perspective of a map?"

ANNOTATED CHILDREN'S LITERATURE FOR TEACHING PERSPECTIVE

Hartman, Gail. *As the Crow Flies: A First Book of Maps*. New York: Aladdin Paperbacks, 1993.

This book uses simple words and pictures to describe different geographical areas from the perspectives of an eagle, a rabbit, a crow, a horse, a gull, and the moon. All maps are joined together to present "The Big Map" at the end of the story, earning this book a four-star rating.

Lauber, Patricia. *Seeing Earth from Space*. New York: Orchard Books, 1990.

Photographs and images of the earth from space, such as the Arctic's ozone hole, Brazilian forest fires, and the long-buried riverbed in the Sahara Desert, help students see the earth from a variety of perspectives. This is a great way to introduce students to remote sensing images.

Spier, Peter. *The Fox Went Out on a Chilly Night*. New York: Dell Dragonfly Books, 1994.

This book presents various panoramic views of a town and countryside as the fox goes out to find supper for his hungry children. The book is based on the old folk song of the same name. The illustrations are not translated into maps but only present a view of various areas from the song.

Taylor, Barbara. *Maps and Mapping*. New York: Kingfisher Books, 1993.

This book begins with a bird's-eye view of a neighborhood and map, then travels to a bedroom. A variety of maps are displayed throughout the book, as well as suggested activities to go along with the maps. Because this book also covers map projections and scale, it can be used to teach various map skills.

ASSESSMENT QUESTIONS—PERSPECTIVE

- What would happen if the earth were a different shape?
  Possible Answer: The weather would change based on the shape. A cube-shaped earth might have temperate weather on the faces and there would be no polar weather. A flat earth may not have gravity. Settlement patterns would shift to different areas. The orbit would also be affected.
- Why is "perspective" important in mapping?
  Possible Answer: Maps reflect ideas of space and place. How a person views a place would determine the perspective. An ant has

a different view of a place than a raven. The intended audience is important to the design of a map.

- What is the difference between an aerial photo and a map? How are they the same?

  Possible Answer: A photograph represents a realistic image that records a scene at a particular instant. A map is a drawing with points and lines that symbolize things in the world. Both can represent the same place.

- What would an aerial photo of your street show?

  Possible Answer: An aerial photo of my street would show a row of houses and condos, fences, and lots of trees. There would also be a sinkhole area.

- Compare a north-oriented map to a south-oriented map.

  Possible Answer: Geographic and religious influences have changed how maps are oriented. North orientation was a reflection of the importance of knowing where the magnetic north was located. Seafaring explorers needed to know how to orient themselves with a compass. Reverse maps appear upside down and are called south-oriented. These maps protested against western hemisphere bias. Zoo or street sign maps are often oriented toward the south to show a tourist where they are.

- Why is a globe more accurate than a map?

  Possible Answer: A globe mimics the true shape of the earth, whereas a flat map presents a distorted view of the earth. Globes provide more accurate shapes, sizes, and locations than flat maps. Flat maps are more flexible and can be folded.

- Why does the size of a continent change on various map projections?

  Possible Answer: Maps must compromise shape, area, distance, or direction. These distortions are what causes the changes on the various projections.

- How would a photograph of our classroom differ from a map of our classroom?

  Possible Answer: A photograph of our classroom would show views from various angles in a realistic manner. A map of our classroom would probably be from an aerial view, and we would draw lines and symbols to represent various items in the room.

- Describe a map of our schoolyard from the view of a snail, a bird, and a child.

  Possible Answer: A snail view would be a very small part of the playground. A bird view would be an aerial view of the whole playground. A child view would be the places that a child notices the most.

NOTES

1. Adapted from lesson "Skinning the Earth," Janice VanCleave, *Geography for Every Kid: Easy Activities That Make Learning Geography Fun* (New York: John Wiley & Sons, Inc., 1993), 48–50.

2. Adapted from ideas by Veronica Getskow, *Incredible Edible Geography* (Irvine, CA: Thomas Bros. Maps Educational Foundation, 1998), 13, 17.

3. "40 Maps That Will Help You Make Sense of the World," Twisted Sifter, accessed November 4, 2019, https://twistedsifter.com/2013/08/maps-that-will-help-you-make-sense-of-the-world/.

4. Adapted from lesson idea by Jeannie Spinella, teacher consultant with the Louisiana Geography Education Alliance (LaGEA).

## CHAPTER 3

Scale

Figure 3.1. *Scale*
Madalyn Stack, artist

The activities in this chapter explore ways to teach the concept of scale. Map scales help us measure distance from one place to another and show relationships between the distance on the map and the distance on the ground. Begin by having students explore synonyms for scale: ratio, proportion, correlation, relationship. Encourage students to become familiar with various types of map scales.

Scales are different on every map. Scale can be represented by words, a graphic bar scale, or a representative fraction or ratio. The most common type of graphic scale looks like a ruler. Also called a bar scale, it is simply a horizontal line marked off in miles, kilometers, or some other unit of measuring distance. The verbal scale is a sentence that relates distance on the map to distance on Earth (e.g., one centimeter represents one kilometer). The representative fraction can be shown as a fraction or ratio—for example, 1/1,000 or 1:1,000. This means that any given unit of measure on the map is equal to one thousand of that unit on Earth. The reader must be able to determine the relationship between a unit of measure on the map and a unit of measure in the real world.

The size of the area covered helps determine the scale of a map. A map that shows an area in great detail, such as a street map of a neighborhood, is called a large-scale map because objects on the map are relatively large and more features are displayed. A map of a larger area, such as a continent or the world, is called a small-scale map because objects on the map are relatively small.

The U.S. Geological Survey has published maps of various scales.[1] Topographic maps in most of the United States have been published at a scale of 1:24,000. Information at this large scale typically includes important buildings, campgrounds, caves, bridges, fence lines, and changes in road lanes. Smaller scale maps (1:250,000) would limit to major features, such as airports, railroads, and major roads.

Typical concepts and skills for learning scale include identifying larger and smaller, miles and kilometers, inch to mile, distance and travel time, and relative and exact distances to places. Activities for learning scale could include mapping the classroom, finding cities located within an inch of one another, and examining a variety of maps at different scales. Students need to be able to understand the relationship between distance on a map and distance in real life.

ACTIVITIES FOR TEACHING SCALE

24. A Scale of Two Cities[2]

Materials: state map, rulers
Time: 25 minutes
Suggested Grade Levels: 4–6

Ask students, "Where can you find two cities in the space of an inch?" Look at a state map and use a ruler to find two cities within an inch of each other. Use the scale to determine the number of miles this distance represents. Find two places on the map that are about the difference of the map scale. Tell students to plan a trip between two cities and determine the miles between the cities based on the map scale.

25. Mapping the Classroom

Materials: large art paper, drawing tools
Time: 30 minutes
Suggested Grade Levels: 3–5

Divide students into two teams and have them make a classroom map to scale. Often the floor tiles are one-foot squares and can be used to create a scale (1 inch = 1 foot). Students need to create a scale and legend on the map. Be sure to include a map key. Students should draw symbols to represent various objects in the classroom.

26. Evaluating Maps

Materials: variety of maps
Time: 30 minutes
Suggested Grade Levels: 4–6

Ask students, "What is a cartographer (a mapmaker)? What do you think a cartographer needs to know?" Introduce the parts of a map: legend (key), compass rose, scale, grid lines, title, and symbols. You can use the acronym DOGSTAILS to remind students what to put on a map (date, orientation, grid, scale, title, author, index, legend, and situation).

Ask students, "How do maps lie?" (They distort reality, stretch some distances and shorten others, use a single symbol to represent many things, misspell place-names, countries fight over what to name places.) Have students examine a variety of maps of the same area and see if they can find any discrepancies. For example, Japan has the "Sea of Japan" but South Korea calls it the "East Sea." Remind students to look at the scale on each map. Does the scale change the perspective of the disputed area?

27. Literature and Maps

Materials: variety of children's books that include maps (e.g., *The Chronicles of Narnia* by C. S. Lewis; *Harry Potter* by J. K. Rowling; *Winnie-the-Pooh* by A. A. Milne)
Time: 45 minutes
Suggested Grade Levels: 3–6

Give the students a variety of children's books that include maps. Determine if the maps are of real places or imaginary places. Determine if the maps are accurate. Determine if the maps are useful. What map elements are missing? Is there a scale included with the map? Have students use the chart in appendix E for evaluating maps in children's books.

28. Tabletop Maps

Materials: tabletop state map from https://www.nationalgeographic.org/topics/state-mapmaker-kits/
Time: 30 minutes
Suggested Grade Levels: 2–4

The National Geographic Society education department has developed a series of tabletop state maps. Print out a tabletop map of your state (you may want to laminate the map to allow students to write on it). The map will be in pieces, so have the students put the map together on a table. Next, have students find several other maps of different sizes of your state. Have students compare the map scales on each.

29. Big to Small

Materials: picture, enlarged picture, reduced picture, document camera, drawing paper
Time: 30 minutes
Suggested Grade Levels: 1–3

Help students understand that an area's size remains constant when it is represented by larger or smaller scales on different maps through an activity involving drawing. Take a picture and enlarge it using a document camera. Trace the drawing on large art paper. Reduce the same picture to a smaller size. Show the students all three drawings. Note that the area in the drawing did not change size in real life, only the representation changed sizes.

30. Blueprints

Materials: blueprints of buildings, such as the school; drawing tools, drawing paper, building blocks
Time: 45 minutes
Suggested Grade Levels: 4–6

Examine the blueprints of your school building. Compare the scale to the actual building size. Have students create their own blueprint of a building at a small scale. Use building blocks to create the building. Reverse the process and build something, then create a blueprint.

31. Shapes to Scales

Materials: construction paper, scissors, three-dimensional objects of shapes
Time: 20 minutes
Suggested Grade Levels: 3–5

Have students cut out various shapes from construction paper (e.g., circle, triangle, square). Compare the cut shapes to three-dimensional objects of a sphere, pyramid, and cube. Talk about concepts such as length, perimeter, and area and how they change when shapes are scaled (enlarged or reduced).

32. Scale Drawing

Materials: cartoon or picture, drawing paper, drawing tools, rulers, scissors, grid paper
Time: 45 minutes
Suggested Grade Levels: 4–6

Give students a picture or a cartoon and tell them to draw the image two times larger. Have students use a ruler and draw grids on the image. Cut the image into squares from the grid. Give each student a piece of the grid to draw. Use a ruler to measure the grid square and multiply times two. Draw the new grid size and complete the drawing at the larger size. Another way to show this is to have two different sizes of grid paper for students to use to draw the same picture. Use something simple, such as a clown.[3]

33. Move the Room

Materials: grid paper, drawing tools
Time: 45 minutes
Suggested Grade Levels: 4–6

Give students grid paper from appendix F and have them draw the classroom. Then, give students another sheet of grid paper and have them "move the room" and draw how the classroom would look a different way. Use the tiles on the classroom floor to help determine the scale of objects in the room. Choose the best room design, and move the classroom to the new design.

34. Jigsaw Puzzle[4]

Materials: maps at two different scales; blank, pre-cut puzzles; Ziploc bags
Time: 45 minutes
Suggested Grade Levels: 4–6

Give students maps at two different scales. Hand out two sets of blank, pre-cut puzzles. Ask students not to tear the puzzles apart. Students must first draw their maps on the puzzle pieces. You may want to let students practice drawing on paper before giving them the puzzles. Students can then color the puzzles. Place the two puzzles in Ziploc bags for storage. Exchange puzzles with other students and figure out how to put the two puzzles together by examining the scale.

ANNOTATED CHILDREN'S LITERATURE FOR TEACHING SCALE

Gonzales, Doreen. *Are We There Yet? Using Map Scales.* North Mankato, MN: Capstone Press, 2007.

Using colorful pictures and a variety of maps, this book teaches students how to understand map scale and figure out how far it is to a place. Definitions of map terms are included throughout the book, along with a few activities.

Rabe, Tish. *There's a Map on My Lap! All about Maps*. New York: Random House, 2002.

In the tradition of *Cat in the Hat* by Dr. Seuss, this book introduces different kinds of maps (city, state, world, topographic, temperature, terrain, and others) and the tools geographers use to read them. The words are in rhyme and include a glossary at the end of the book. Scale, legend, latitude, longitude, compass rose, and grid are illustrated, with rhymed explanations simple enough for the primary child to grasp the concepts.

Wade, Mary Dodson. *Map Scales* (Rookie Read-about Geography series). New York: Scholastic, 2012.

This book teaches students how scales measure distances. The photographs and drawings are colorful, and the text is simple. This series of read-aloud books makes learning about geography fun for younger children.

ASSESSMENT QUESTIONS—SCALE

- Why do you need to know about map scale?
  Possible Answer: Scales can be written as ratios, bars, or words. Scales are the distance on a map in a corresponding ratio of distance on the ground.

- What is the purpose of scale?
  Possible Answer: Scales give you an understanding of distance. If the scale on a map states 1:1,000, that would mean that 1 centimeter on the map would equal 1,000 kilometers on the ground.

- What is the difference between a map scale and a map legend?
  Possible Answer: A scale on a map indicates the measurements of distance between points on a map. A legend indicates symbols that represent a place on a map.

- What is a cartographer? How does a cartographer determine map scales?
  Possible Answer: A cartographer is a person who creates maps. The map scale depends on the purpose of the map. The cartographer will determine whether to use a graphic scale, a verbal scale, or a representative fraction.

- Why do some books have maps?
  Possible Answer: A map can help you understand a story. It reveals the journeys of the characters and the amazing places they visit. It can explain territories and show challenges that must be faced. A map can create a world that doesn't exist.
- Why do countries fight over the names of places on a map? (e.g., to South Korea it is the East Sea, but to Japan it is the Sea of Japan.)
  Possible Answer: Countries often want to claim the same place. By choosing a name that reflects their ownership or culture, this adds substance to their claims. Both Korea and Japan have offered proof of ownership through old maps. The territory would allow shipping and fishing rights to the country that owns it.
- Compare a blueprint to a map.
  Possible Answer: A blueprint is a technical drawing of a room or building that is drawn to scale. Although you can draw a map of a room or building, a map can also include the grounds around the object.
- How would you determine the map scale for a city? a park?
  Possible Answer: You would need to start with the measurement of the city or the park in miles, yards, or feet. This would help you determine the scale for the map. You want the map to fit onto the paper. The intended use of the map would also help determine the scale. A larger scale would show more features.
- What kinds of map scales exist?
  Possible Answer: Verbal scales express measurement with words: 1 inch represents 100 miles. Bar scales show a diagram with the inches measured out. Ratio scales write out the scale in this form: 1:100.

NOTES

1. "Map Scales," U.S. Department of the Interior, U.S. Geological Survey, Earth Science Information Center (ESIC), accessed November 5, 2019, https://pubs.usgs.gov/unnumbered/70039582/report.pdf.

2. Adapted from lesson "Measuring Distances on a Map." National Geographic Education, accessed November 4, 2019, https://www.nationalgeographic.org/activity/measuring-distances-map/.

3. Adapted from lesson "Enlarger," Janice VanCleave, *Geography for Every Kid: Easy Activities That Make Learning Geography Fun* (New York: John Wiley & Sons, Inc., 1993), 64–66.

4. Adapted from lesson by Peggy Meaux, teacher consultant with the Louisiana Geography Education Alliance (LaGEA).

## CHAPTER 4

Orientation

Figure 4.1. *Orientation*
Madalyn Stack, artist

The activities in this chapter explore ways to teach the concept of orientation. Orientation indicates direction and is commonly represented by a north arrow or compass rose. Begin instruction with students by exploring synonyms for orientation: positioning, location, situation, bearings, placement, direction, alignment. Typical concepts and skills taught in schools for orientation include directional terms (e.g., left, right, down, north, south, east, west), cardinal and intermediate directions, parallels and meridians, and systematic orientations of maps.

The orientation of a map should indicate which way is north, south, east, and west. By convention, north is toward the top of the page; however, north does not have to be at this location. Religious and geographic influences over time affected the orientation of maps. Early mapmakers in Europe drew the *Orient* (China) at the top, from which we get the idea to "orient" a map. On the earth, true north (the direction to the North Pole) differs from magnetic north, and the magnetic north pole moves due to changing geophysical conditions of the earth's crust and core. Many reference maps indicate both.

A compass rose is a figure on a compass, map, nautical chart, or monument used to display the orientation of the cardinal directions (north, east, south, and west) and their intermediate points (northeast, northwest, southeast, and southwest). Formerly known as the "wind rose" (indicating the directions of the winds), the device has thirty-two points to note the eight major winds, eight half winds, and sixteen quarter winds.[1]

Students should know that the equator is located at 0 degrees, the North Pole is 90 degrees north of the equator, and the South Pole is 90 degrees south of the equator. The prime meridian is 0 degrees longitude, and the international date line is 180 degrees. Each fifteen degrees for the longitude lines represents a one-hour rotation of Earth. Greenwich Meridian was chosen as the prime meridian of the world in 1884 due to the great expansion of railways and communication networks. Forty-one delegates from twenty-five nations met to determine this line.[2] The line itself divided the eastern and western hemispheres of the earth—just as the equator divides the northern and southern hemispheres.

Without the international date line, people who travel west around the planet would discover that when they returned home, it would seem as though an extra day had passed. Countries are on either side of the international date line, which runs down the middle of the Pacific Ocean. If you cross the date line moving east, you subtract a day. But if you are moving west you add a day.

Before the nineteenth century, almost every town in the world kept its own local time. There were no laws that set how time should be measured, when the day would begin and end, or what length an hour might be. A Canadian engineer and inventor named Sandford Fleming suggested that the earth should be divided into twenty-four time zones, since it takes the earth twenty-four hours to make one rotation.[3] Each time zone is about fifteen degrees of longitude wide, with clocks to the east one hour later and clocks to the west one hour earlier.

Activities for learning orientation can include hunting for pirate treasure, building a compass rose, and reading a compass. Many students already know how to use a phone to find places and the direction in which to travel. They may not be familiar with the term "orientation."

ACTIVITIES FOR TEACHING ORIENTATION

35. Cardinal Directions

Materials: U.S. map, signs to indicate direction, masking tape, compass
Time: 30 minutes
Suggested Grade Levels: K–2

Explain to the students the cardinal directions—north, east, south, and west. Use the mnemonic Never Eat Soggy Waffles. Introduce the intermediate directions—northeast, northwest, southeast, and southwest. Display a U.S. map, then ask students, "In which direction would you travel from our state to get to Washington, D.C.? to Los Angeles? to Houston?" Create signs for each direction, use a compass to determine directions, and then tape the signs on the walls around the classroom. Ask students, "Which direction would you travel to get out the door? to talk to the teacher? to sharpen your pencil?"

36. Orientation Games

Materials: Twister game
Time: 20 minutes
Suggested Grade Levels: K–2

Games such as Simon Says, the "Hokey Pokey," and Twister can be used to teach orientation skills. Have students move body parts from left to right, over or under, above or below, around or through, beside or between something. You could use the game I Spy to have students find things in the classroom (e.g., I spy with my little eye something on the west side of the classroom). Use directional terms in everyday language. For example, say things like, "Put your shoes next to the door," or "At the end of this hall, we are going to turn left."

37. Pirate Treasure Hunt[4]

Materials: pirate hat, eye patch, atlases for each group of students, clues, answer sheet, treasure box with prizes
Time: 45 minutes
Suggested Grade Levels: 3–5

Wear a pirate hat and eye patch to class. Tell students that you have intercepted and decoded clues that will lead to a hidden pirate treasure. They will compete in teams to follow the clues and win the treasure. Students will need their skills in orientation and reading an atlas to find the treasure. Divide the class into groups of four. Students need a navigator to read an atlas, a reader to read aloud each clue to the group, a writer to write down the answer on the clue answer sheet, and a runner to take the answer to the teacher to check for accuracy and receive the next clue. The first team to get all the clues wins the treasure (e.g., a decorated box with beads and coins).

Example clues: (1) Your search for the buried treasure begins near Los Angeles. Name the islands located nearby. (2) Here at the Channel Islands, you meet an old sailor who directs you on the next part of the journey to Hawaii. What is the elevation in feet of the Mauna Loa volcano? (3) In a cave near the volcano, you find the treasure map. Name the gulf that lies 43 degrees north, 68 degrees west. (4) It is rather cold at the gulf, and you don't want to stay long. You pick up a key to the treasure box and head south. Name the large city south of Lake Okeechobee. (5) Watch out for those alligators! Now if you hurry, you can be the first team to get to the treasure. Name the latitude and longitude of New Orleans, LA.

Answers for clues: (1) Channel Islands; (2) 13,678 feet; (3) Gulf of Maine; (4) Miami, FL; (5) 30 degrees N, 90 degrees W

38. Reading a Compass

Materials: compass for each student, drawing paper and drawing tools, treasure
Time: 30 minutes
Suggested Grade Levels: 2–6

Read the article at the REI website on how to use a compass (https:// www.rei.com/learn/expert-advice/navigation-basics.html). Give each student a compass and provide the following instructions:

- Hold the compass flat so that the needle rotates freely.
- Note that the red needle points toward Earth's magnetic north.
- Turn the bezel to the direction you wish to go.
- Put "red" in the "shed" (drawn arrow) to align properly (by moving your body).
- Follow the pointer.

Next, give directions to students to follow. For example, walk ten steps to the east, four steps to the north, and six steps to the west. Which direction do you need to walk to go back to your seat? How many steps will it take to arrive? Take the students outside and let them hide a "treasure" to find and create a map with clues that require a compass.

39. Hemispheres

Materials: clay, plastic knives
Time: 15 minutes
Suggested Grade Levels: K–2

Give students a ball of clay and a plastic knife. Ask students to show all the ways the clay could be cut in half. Geographers divide the earth into halves called hemispheres. The equator divides the northern and southern hemispheres, while the prime meridian divides the western and eastern hemispheres.

40. Time Zones

Materials: book *Somewhere in the World Right Now* (Schuett, 1997), book *Is Anybody Up?* (Kandoian, 1989)
Time: 30 minutes
Suggested Grade Levels: 4–6

Show students the time zone map in the book *Somewhere in the World Right Now*. The book includes an explanation of the system of standard times, as it shows what people around the world are doing as you go to sleep. Students could identify regions or hemispheres, as well as the time zone for each image as you read the story. Have students examine political and geographical boundaries in one country (or state) and how they affect the selection of the time zones. For example, the state of Kentucky has both the Central and Eastern Time zones. The move to Eastern Time was an effort to attract commerce from larger cities on Eastern Time.

Another book to use would be *Is Anybody Up?* Molly is making breakfast as the book shows what others are doing around the world—her grandfather in Miami, a boy in Peru, an Inuit woman on Baffin Bay in the North Atlantic Ocean. They are all eating breakfast because they are all in the same time zone.

41. What Time Is It?[5]

Materials: book *Longitude: The True Story of a Lone Genius Who Solved the Greatest Scientific Problem of His Time* (Sobel, 1995), wall clocks
Time: 30 minutes
Suggested Grade Levels: 4–6

Read excerpts from *Longitude: The True Story of a Lone Genius Who Solved the Greatest Scientific Problem of His Time*. Put up a wall of clocks in the classroom to show time zones around the world. Arrange the clocks geographically according to longitude. Talk to students about what might be going on in other places in the world during various times of the school day. Ask students questions, such as, "How would your day be different without clocks? What if the world had no time zones?"

42. Upside-Down World

Materials: map with New Zealand at the top
Time: 20 minutes
Suggested Grade Levels: 2–4

Show students a map in which New Zealand is at the top. The curiosity of students will be aroused as they ponder such questions as, "Is it 'correct' to show a map south-up? Why is the world always shown north-up? Are there other ways to view the world?" Some things that students may notice on the upside-down map is that Russia and the oceans look very big, while Europe looks tiny. South America is further east, which is noticeable on the north-oriented map.

43. Which Way Is North?

Materials: compass, GPS unit, computer with Google Earth, grid
    paper
Time: 45 minutes
Suggested Grade Levels: 4–6

Take students out onto the playground. Ask them to look for the sun and determine which direction their school is facing. Students can record the position of the sun in the morning and afternoon. Use a compass to accurately assess the direction the school faces. Use a GPS to take latitude and longitude numbers at different points around the playground. Look up your school location with Google Earth. Have students use the grid paper in appendix F to map the playground using compass directions and GPS locations.

44. Building a Compass Rose

Materials: Hula-Hoop; two yardsticks; poster labels N, S, E, W
Time: 30 minutes
Suggested Grade Levels: K–2

Place a Hula-Hoop on the floor in the middle of the classroom. Use two yardsticks to make a plus sign in the middle of the hoop. Label N, S, W, and E on poster board and place the labels on the tips of the yardsticks. Have students walk to the west side of the room, hop to the east side, run to the north, and walk backward to the south. If you have a class pet or stuffed animal, have students move the animal using cardinal directions.

45. Use Your Brain to Navigate[6]

Materials: hiking map, pencil, drawing paper, GPS units or phone apps, book *Finding Your Way without Map or Compass* (Gatty, 1998)
Time: 1 day
Suggested Grade Levels: 5–6

The hippocampus part of the brain helps us navigate the spatial environment. The use of smartphones and other navigational technology has created a loss in the ability of hikers to navigate their way in unfamiliar terrain. Read excerpts of *Finding Your Way without Map or Compass* to students. Take students out to a safe area that is good for hiking. Divide the students into three groups. Give one group of students a map to follow to reach a central spot. Have students note landmarks along the way. Give general directions to another group of students (but no map) and ask them to reach the same spot. Also ask them to note specific landmarks they find along the way. The third group of students are allowed to use GPS or phone apps on their hike. When the groups arrive at the designated spots, compare the time it took each group to arrive. Have each group draw a map of their journey and compare landmarks. Discuss navigation and orienting skills with students.

ANNOTATED CHILDREN'S LITERATURE FOR TEACHING ORIENTATION

Gatty, Harold. *Finding Your Way without Map or Compass.* Mineola, NY: Dover Publications, 1998.
Written by one of the world's great navigators, this book shows how to find your way in the wilderness by studying wind directions, noticing reflections in the sky, observing animals and weather patterns, and even using smell and hearing. Pathfinders can learn through this book how to use the signs in the natural world to find their way.

Greve, Meg. *North, South, East, and West.* Vero Beach, FL: Rourke Educational Media, 2009.
Simple text and photos are used to explain directional terms. The book includes a glossary of terms and is written on the level for a kindergartener or first grade student.

Kandoian, Ellen. *Is Anybody Up?* New York: Putnam Juvenile, 1989.
As Molly is making breakfast, the book shows what others are doing around the world—a sailor off the coast of Chile, a boy in Quebec, and a cat in New York. They are all eating breakfast because they are all in the same time zone.

Schuett, Stacey. *Somewhere in the World Right Now.* New York: Dragonfly Books, 1997.
This book includes a map of time zones. The story describes how when one person is going to sleep, other people are waking up and starting a new day.

Sobel, Dava. *Longitude: The True Story of a Lone Genius Who Solved the Greatest Scientific Problem of His Time.* London: Walker Books, 1995.
This book describes the forty-year quest of an uneducated clockmaker to solve the problem of how to measure longitude while at sea. Responding to

a nationally sponsored contest, John Harrison spent his life developing what came to be known as the chronometer.

ASSESSMENT QUESTIONS—ORIENTATION

- What are cardinal directions?
  Answer: The four cardinal directions are north, south, east, and west.
- What are intermediate directions?
  Answer: The intermediate directions are between the cardinal directions. They are northeast, northwest, southeast, and southwest.
- Why are directions important?
  Possible Answer: Directions help us orient where we are and where we want to go.
- What is the difference between magnetic north and true north?
  Possible Answer: A compass needle points to the magnetic north, which actually moves every year. The geographic North Pole is where lines of longitude converge.
- What is the difference between latitude and longitude?
  Possible Answer: A latitude line measures distance north or south of the equator. A longitude line measures distance east or west of the prime meridian.
- What is the purpose of the international date line?
  Possible Answer: The international date line serves as a line of demarcation separating two consecutive calendar dates. If you cross it westward, the day goes forward by one. If you move in an easterly direction, you must subtract a day.
- Compare a map with New Zealand at the top with a typical world map.
  Possible Answer: On a north-oriented map, New Zealand looks much smaller and less significant. By being at the top of the world on a south-oriented map, New Zealand looks larger and seems more significant.
- What is the difference between a compass and a GPS unit?
  Possible Answer: A compass is a simple navigation tool that indicates the direction of magnetic north. It is inexpensive and doesn't require a power source. A GPS device is a receiver of information

from the twenty-four satellites that orbit the earth used to compute location. A GPS device can track your movement and store maps. A GPS needs a clear area to be able to pick up a satellite signal.

- How would you get from the front office to the playground?
  Possible Answer: To get from the office to the playground, I need to walk down the hall and turn left in the blue hall. After I pass the library, I turn right down the short green hallway to the doors that lead to the playground.
- What are some orienteering or navigation tools?
  Possible Answer: Some orienteering tools include a map, compass, and a GPS device.
- Name four items placed on the east side of the classroom.
  Possible Answer: The teacher's desk, a bookshelf, a computer, and a garbage can are on the east side of the room.
- When it is noon in Paris, what time is it in San Francisco?
  Answer: 3:00 a.m. in San Francisco of the same day
- If it is 3:00 p.m. in New York, what time is it in London?
  Answer: 8:00 p.m. in London of the same day
- Why do we have time zones?
  Possible Answer: Since different parts of the earth receive sunlight or darkness, we need different time zones.

NOTES

1. Bill Thoen, "Origins of the Compass Rose," GISnet, accessed November 4, 2019, http://www.gisnet.com/notebook/comprose.php.

2. "What Is the Prime Meridian and Why Is It in Greenwich?" Royal Museums Greenwich, accessed November 1, 2019, http://www.rmg.co.uk/discover/explore/prime-meridian-greenwich.

3. Janice VanCleave, *Geography for Every Kid: Easy Activities That Make Learning Geography Fun* (New York: John Wiley & Sons, Inc., 1993), 137.

4. Adapted from lesson idea by Dianne McWilliams, teacher consultant with the Louisiana Geography Education Alliance (LaGEA).

5. Adapted from lesson "World Clocks: Exploring Local Time and Time Zones." Journey North, accessed November 4, 2019, https://journeynorth.org/tm/mclass/WorldClocks.html.

6. Adapted from lesson ideas by Rebecca Maxwell, "Spatial Orientation and the Brain: The Effects of Map Reading and Navigation," March 8, 2013, GIS Lounge, accessed November 4, 2019. https://www.gislounge.com/spatial-orientation-and-the-brain-the-effects-of-map-reading-and-navigation/.

## CHAPTER 5

Map Symbols and Map Keys

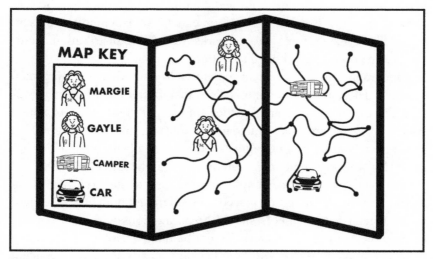

Figure 5.1. *Map Keys and Symbols*
Madalyn Stack, artist

Map symbols and colors are used to represent real objects to ensure that a person can read a map. The symbols are located in boxes in the corner of a map called the map key or map legend. A map key is essential to understanding the meaning of what the map represents. Symbols may be pictures or icons, such as a triangle for a mountain or an oil derrick to show where oil is found. Often, the map may be colored or shaded, so the map key explains what the colors and shades mean. Map keys can be found on all types of maps: highway, subway, treasure, and even video game maps.

Begin by having students share synonyms for legend and key: caption, inscription, code, explanation, cipher, guide, pointer, cue, answer. Discuss how a map key or legend can give an explanation and guide the user to the proper use of a map. Share synonyms for symbol: sign, character, mark, hieroglyph, token, emblem, figure, image. Map skills for the elementary years include having students recognize pictorial symbols for familiar features, colors for land and water, symbols for towns and cities, uses of shading, dots and contour lines, and symbols on special features maps to show relief, elevation, and other features.

There are three types of map symbols: point, line, and area. Point symbols include things like individual signs and dots to represent man-made structures, place, or positional data. Line symbols may represent a road, railway, boundary, river, or canal. Area symbols represent things like a marsh, forest, water, parks, or wildlife distribution. Color can also be used to symbolize something on a map.

Topographic maps are great examples to use when teaching map symbols. These maps use contour lines that join points of equal elevation. The closer the lines are together, the steeper the slope represented. Bodies of water are blue, boundaries are black, and roads are red on topo maps. Houses may be represented as small black squares. Further symbols for topographic maps can be found at the USGS website.[1]

Map symbols should be simple and clear and include relevant information.[2] Symbols should be placed on the left and defined to the right in the legend box. Symbols should be harmonious with the map and relate directly to the characteristics on the map. There are several key ideas about symbols that will help teachers clearly define the concept to students.[3]

- Symbols are abstract.
- Symbols represent real objects, whether tangible or intangible.
- Symbols stand for relations among objects.
- Each symbol establishes the existence and location of the object that it stands for.
- The symbols on a map represent choices by the mapmaker.

Students often misinterpret symbols on maps. Students often get confused that the size of the dot or circle indicates a larger or smaller city, which it often does not.[4] The green color for low elevation may be misunderstood as a forest or grassland. A pictorial symbol may be misunderstood as only one corn stalk or one car exists in that place.

As children learn to interpret then create symbols, they move from concrete to abstract understanding.[5] Children need firsthand experiences in order to realize that a picture or icon can represent a real object. Activities for teaching map symbols and map keys can include board games, construction of familiar areas, comparing and contrasting map keys, and clay models of symbols.

ACTIVITIES FOR TEACHING MAP SYMBOLS AND MAP KEYS

46. What Is It?

Materials: pictures of traffic signs, map with simple symbols and features, photographs of landforms, symbol cards, drawing paper, drawing tools
Time: 30 minutes
Suggested Grade Levels: K–1

Show students pictures of typical traffic signs (e.g., stop sign, yield, hospital zone, walk, airport). Talk about how these "symbols" represent ideas or places. Introduce the concept of map symbols. Play a game with students in which you name a feature on the map and ask the students to point to the symbol they think represents that feature. Show students photographs of various landforms (e.g., lake, mountain, island) and have them match the symbol from the cards to the landform it might represent. Ask students to draw symbols to represent objects in the classroom.

47. Map Symbol Bingo

Materials: Bingo card, drawing tools, various maps with map keys
Time: 30 minutes
Suggested Grade Levels: 2–4

Show students a variety of map keys and note the various map symbols represented. Then give students the blank Bingo card from appendix G and have them draw in at least ten map symbols that they remember seeing. Call out or show various symbols and have the students put an X or a Bingo marker on the symbols you call. The student with the most symbols covered wins the game.

48. Creating Symbols

Materials: books *Keys and Symbols on Maps* (Greve, 2009) and
  Map Keys (Olien, 2012)
Time: 30 minutes
Suggested Grade Levels: 1–3

Read to students the books *Keys and Symbols on Maps* and *Map Keys*. Discuss various symbols that are displayed in the books and how they represent real objects. Give each student a ball of clay and assign various symbols to create. You might demonstrate a 3D model of a mountain, then show a 2D model of a triangle to represent a mountain in order to explain how a symbol might differ from a model. Allow students to create symbols, then ask another student to guess what their symbol represents.

49. A Symbol by Any Other Name

Materials: variety of maps with map keys
Time: 45 minutes
Suggested Grade Levels: 4–6

There is little agreement about and no standardization for map symbols. Color schemes vary from browns to greenish brown or reddish brown to denote high elevations, and shades of green and blue to denote land that is below sea level. Have students examine map symbols from a variety of maps. How are they similar? What differences do you see? Do the keys agree with the maps? Are the symbol orders the same (facing upward, downward, left, or right)?[6]

The same symbol can often stand for different information (e.g., a triangle could represent a mountain peak or a historical marker). Have students find two different maps that use different symbols to represent the same piece of information. Then have students find two different maps that use the same symbol to represent different pieces of information.[7]

50. Neighborhood Symbols[8]

Materials: masking tape, clay or construction paper, large building blocks
Time: 1 hour
Suggested Grade Levels: 2–4

Remind students that real objects can be represented by symbols. Have students recall objects and places that they have seen and discuss how these might be represented as a symbol. Take students outside to visit a place (e.g., playground, park). When returning to the classroom, have students use building blocks to reconstruct the area they visited. Encourage children to walk through their map area and discuss the things they saw. Create three-dimensional figures of trees or park benches to add realistic views of the map.

Create a map of the neighborhood using masking tape to represent streets. Identify and label the significant streets. Have students create clay models or construction paper models to represent their homes. Place the houses on the correct streets of the neighborhood. Have students give specific directions to their home from the school.

51. Matching Symbols[9]

Materials: teacher-made outline maps of classroom or play-
  ground, cut-out symbols of construction paper
Time: 30 minutes
Suggested Grade Levels: K–1

Provide a teacher-made outline map of the classroom or playground
for each student. Cut out symbols from construction paper of various
objects in these environments and have the students put the symbols
on the map. Create a large map legend on the board with the symbols
used on the maps and discuss the importance of reading a map legend
to understand a map.

52. Points, Lines, and Areas[10]

Materials: variety of maps with symbols represented by points,
  lines, and areas
Time: 25 minutes
Suggested Grade Levels: 2–4

Choose a map with lots of symbols on it. Let the reader hold the map
so that the other person cannot see it. The reader will then read symbols
from the map. The other person will state whether he or she thinks that
symbol is represented by a point, a line, or an area. Check to see if the
guess was accurate, then switch sides.

53. Game On!

Materials: Monopoly or other board game pieces
Time: 30 minutes
Suggested Grade Levels: 4–6

Games, such as Monopoly, have tokens and symbols that can be used to reinforce the idea of symbols representing an object. Some Monopoly pieces include top hat, iron, thimble, boot, battleship, race car, dog, cat, wheelbarrow, and sack of money. Have students look around the classroom and see what objects might make good game pieces (e.g., eraser, button, push pin). Have the students design symbols for a game board based on items they find. You might introduce how an icon is a graphic symbol. Have students examine various games and discuss the symbols or icons they find. Are some symbols difficult to decipher? What might be a better symbol?

54. A Legend in Its Own Time

Materials: variety of maps with no legends, drawing paper, drawing tools
Time: 30 minutes
Suggested Grade Levels: 4–6

Must a map have a legend? Find a variety of maps that have no legends, and ask students to determine what the maps might represent. Then ask students to create legends for the maps. How do they decide which symbols to include? Will they use points, lines, or color for the symbols? Reverse the process and start with the legend, then create a map.

55. Cookie Symbols[11]

Materials: graham crackers, frosting, plastic knives, paper towels, M&M candies
Time: 30 minutes
Suggested Grade Levels: 1–2

Give each student a graham cracker and allow the students to spread icing to cover the cookie. Explain to the students that the graham cracker represents the classroom. Assign a few items for students to place on the cookie, such as asking students to put a yellow M&M where the teacher's desk is in the classroom or put a green M&M where the door is. Ask the students to choose a color for themselves and place on the cookie their approximate location in the classroom. Talk about how the candies are symbols to represent items in the classroom, just as maps have symbols to represent items in the real world.

56. Mapping Mnemonics

Materials: none
Time: 25 minutes
Suggested Grade Levels: 3–6

Mnemonics are memory-aid devices that can be used for information retention. Elementary children often learn *Never Eat Soggy Waffles* or *Never Eat Slimy Worms* to memorize the compass directions (North, East, South, West), or *HOMES* to represent the Great Lakes (Huron, Ontario, Michigan, Erie, Superior). *TOADS on LOGS* can be used to remember all the parts of a map: title, orientation, author, date, scale, legend, outline, grid, and source. To remember the countries in Central America, use the sayings, *By Gosh Eating Spicy Hot Nachos Causes Real Pain* or *Big Gorillas Eat Hotdogs, Not Cold Pizza* (Belize, Guatemala, El Salvador, Honduras, Nicaragua, Costa Rica, Panama). Now, challenge students to make their own mnemonics to memorize the countries in Africa!

ANNOTATED CHILDREN'S LITERATURE FOR TEACHING MAP SYMBOLS AND MAP KEYS

Boothroyd, Jennifer. *Map My Neighborhood* (First Step Nonfiction—Map It Out). Minneapolis: LernerClassroom, 2013.

Color photographs and drawings help young readers learn about maps and map keys. A young girl maps out her neighborhood for her visiting grandmother. Instructions are included for making your own neighborhood map.

Greve, Meg. *Keys and Symbols on Maps* (Little World Geography). Vero Beach, FL: Rourke Educational Media, 2009.

Younger students can learn about keys and symbols on maps through colorful photos and simple text. A glossary of terms is included.

Olien, Rebecca. *Map Keys* (Rookie Read-about Geography series). New York: Scholastic, 2012.

Another in the read-aloud geography series, this book introduces the importance of having map keys and how to use them. Younger children will enjoy this introduction to parts of a map. Topics include reading a map, pictures and shapes, symbols, colors and lines, kinds of maps, and making maps.

ASSESSMENT QUESTIONS—MAP SYMBOLS AND MAP KEYS

- What is the purpose of a map key?
  Possible Answer: A map key explains what the symbols or colors on a map mean.
- Using the map key, find the post office and the picnic table on the map.
  Possible Answer: (Teacher provides a map that includes simple symbols of a post office and picnic table.)
- Who do you think would look for symbols that show a car park, a campsite, and an information point?
  Possible Answer: A tourist or person on a vacation might look for a car park, campsite, and information point.
- You are lost in the woods and need to find water. Look at the map and find water sources. Plot your route to the nearest water source.
  Possible Answer: (Teacher provides map with a hiking trail and water sources.)
- Draw two different ways to represent a house on a map.
  Possible Answer: A house could be in the shape of a square or in the shape of a square with a triangle roof.
- How many picnic tables are in the park?
  Possible Answer: (Teacher provides map of a park that includes symbols of picnic tables.)

- Follow the forest path through the zoo. What animals do you find along the way?
  Possible Answer: (Teacher provides map of a zoo.) If I follow the forest path on the map of the zoo, I find three animals along the way: the giraffe, the ape, and the gazelle.
- What symbols might you create on a map of your neighborhood?
  Possible Answer: For my neighborhood, I would create a symbol for houses, a symbol for light poles, a symbol for fire hydrants, and a symbol for mailboxes.
- What is the symbol for capital cities?
  Possible Answer: A capital city is often marked with a star.

NOTES

1. "Topographic Map Symbols," U.S. Geological Survey (USGS), accessed November 4, 2019, https://pubs.usgs.gov/gip/TopographicMapSymbols/topomapsymbols.pdf.

2. Jason Dykes, Jo Wood, and Aiden Slingsby. "Rethinking Map Legends with Visualization," *IEEE Transactions on Visualization and Computer Graphics* 16, no. 6 (2010): 890–99.

3. Madeleine Gregg, "Seven Journeys to Map Symbols: Multiple Intelligences Applied to Map Learning," *Journal of Geography* 96, no. 3 (1997): 146–52.

4. Sarah Witham Bednarz, Gillian Acheson, and Robert S. Bednarz, "Maps and Map Learning in Social Studies," *Social Education* 70, no. 7 (2006): 398–404, 432.

5. Betty Richards, "Mapping: An Introduction to Symbols," *Young Children* 31, no. 2 (1976): 145–56.

6. Lesson idea from David J. Swartz, "A Study of Variation in Map Symbols," *The Elementary School Journal* 33, no. 9 (1933): 678–79.

7. Lesson idea from Gregg, "Seven Journeys to Map Symbols," 148.

8. Lesson idea from Richards, "Mapping: An Introduction to Symbols," 147.

9. Lesson idea from Bruce M. Frazee, "Foundations for an Elementary Map Skills Program," *The Social Studies* 75, no. 2 (1984): 79–82.

10. Lesson idea from Gregg, "Seven Journeys to Map Symbols," 148.

11. Lesson idea from Melinda Schoenfeldt, "Geographic Literacy and Young Learners," *The Educational Forum* 66, no. 1 (2001): 26–31, DOI: 10.1080/00131720108984796.

## CHAPTER 6

Types of Maps

Figure 6.1. *Types of Maps*
Madalyn Stack, artist

A good definition for a map is a representation of an area on a flat surface. Begin instruction with students by exploring synonyms for map: chart, plot, draw, depict, portray. Early maps were pictorial in nature and began with lines drawn in dirt or etched into clay, painted on rock walls or animal skins, carved into wood, or made of sticks and shells. From delicate silk maps in China to simple sketches on clay tablets by the Babylonians, early maps had no rules as to how they were oriented. They often depicted small areas, such as a city, a trade route, a hunting ground, or a military campaign. Maps were used for navigation across unknown seas or land, as routes for religious pilgrims, or as a record of ownership.

Greeks and Romans led the way in early mapmaking. A Roman (some sources say "Greek") geographer named Ptolemy published a book of maps around 151 CE that featured lines of latitude and longitude. In the Middle Ages in Europe, most maps were produced within monasteries. Jerusalem was often placed in the center of maps, with angelic cherubs and sea monsters decorating the fringes. It was the Islamic world that developed more scientific mapping, specifically through an Arab scholar named Al-Idrisi who produced world maps and geography books.[1]

The invention of the printing press, the discovery of the Americas, and a general thirst for knowledge led to an expansion of geographic knowledge during the Renaissance period. Today, modern satellite systems, GPS (global positioning system), GIS (geographic information system), and remote sensing contribute to accurate mapping techniques to produce a variety of types of maps. Internet tools allow students to become cartographers and create their own maps.

There are several basic issues that cartographers must consider in map design. One of the first steps is to consider the purpose and audience for the map. This will determine the essential map elements and layout of the map. Next, the cartographer should choose the map type (reference or thematic). Reference maps depict selected details of the physical and human environment. Google Maps are classic examples of reference maps, which are mostly intended to show where things are. Thematic maps depict spatial patterns of selected data. The map made by London physician John Snow to show the cholera outbreak is an example of a thematic map. Selecting a good title, deciding where to place text in relation to map features, and designing an overall layout that is informative and pleasing are important considerations for map design.

Activities to teach types of maps can include everything from puzzles to cakes, pirates to play dough. There are so many different kinds of maps for students to explore, including treasure, topographic, hiking, malls, political, physical, climate, economic, under sea, and even space. Students should process maps from large to small, from real to imaginary, from sculpted to drawn. Critical thinking and inquiry skills are promoted through the use of maps in the elementary classroom.

ACTIVITIES FOR TEACHING TYPES OF MAPS

57. Map Bingo

Materials: Bingo grid, PowerPoint of pictures of various types
of maps
Time: 25 minutes
Suggested Grade Levels: 3–6

Using the Bingo card from appendix G, have students list different
kinds of maps that they know on the squares (e.g., zoo map, weather
map, map of rivers, road map, state map, historical map, campus map,
city map, population map, topographic map, railroad map, treasure
map, map of regions, time zones, product map, airport terminal, ball-
park seating, elevation map, ocean floor map, fire drill map, and mall
map).

Develop a PowerPoint with at least thirty kinds of maps. Show the
PowerPoint to students and have the students put an X on the map
list they created to match maps that they see on the PowerPoint. The
student who covers all maps on their paper wins. Ask students, "How
many maps from the PowerPoint and your Bingo card were the same?
How many were different? What maps do you have that were not listed
on the PowerPoint? What did you discover about the types of maps?"
You may be surprised at the types of maps the students list on their
grids.

58. Map Your Land[2]

Materials: long sheet of butcher paper, markers, cut pieces of
construction paper in green and black, personal items (e.g.,
watch, ring, barrette)
Time: 45 minutes
Suggested Grade Levels: 4–6

Draw a river through the center of the sheet of butcher paper with a lake at one end. Assign sections of property along the riverbank to students and give them the opportunity to determine if they will use the land for agriculture, recreation, or business purposes. Have students draw their plans from an aerial viewpoint. Cut small pieces of colored paper to represent different sources of pollution (e.g., black for oil run-off from parking lots; green for nitrogen from lawn fertilizer).

As each student explains how they have chosen to use their property, the class decides what impact the property owners have on the river and add small amounts of colored paper to represent pollution. Ask each student to place a personal item (e.g., watch, pen, or ring) in the river and wash everything down to the lake. Students will then see that some pollutants (personal items) can be traced back to an exact point and other pollutants cannot be traced back to one point. Have the students examine their own state's river systems and determine what human influences might affect the environment.

59. Satellite Maps[3]

Materials: satellite images of the U.S. at night (from website https://www.nasa.gov/mission_pages/NPP/news/earth-at-night.html) laminated and cut into puzzle pieces, Ziploc bags
Time: 30 minutes
Suggested Grade Levels: 4–6

Print out the satellite images of the U.S. at night. Laminate and cut the images into puzzle pieces and place into Ziploc bags. Assign each student a partner to put together a puzzle. Do not tell the students what image is on the puzzle. Once students put the puzzles together, ask them to identify the type of map (satellite map of the U.S.). Ask students to find the major interstate lines by following the lines of light. Find the major cities (e.g., Los Angeles; Dallas; Washington, D.C.). Ask students, "What does the absence of light mean? (Point out areas that are uninhabited.) Why do you see lights in the ocean? (Oil rigs) Why are there more lights on the eastern side of the U.S. than the western side? (More mountains and deserts are uninhabitable.)"

60. Cartograms[4]

Materials: grid paper, markers, World Population Map, example cartogram map, website for population numbers (https://www. worldometers.info/world-population/population-by-country/)
Time: 45 minutes
Suggested Grade Levels: 4–6

Cartograms are used to shock the reader with unexpected spatial peculiarities: size is proportional to data. Some examples include election results and representations of population. Give students a sheet of grid paper from appendix F and tell them that each square will represent 25 million people. Students will research the population of major continents or specific countries and color-code the squares to represent the population of the assigned areas (e.g., Europe will be represented as about thirty blue squares). Then have students examine a World Population Map and compare their representations to the map.

61. Choropleth Maps

Materials: drawing paper, drawing tools, crayons, choropleth example map
Time: 30 minutes
Suggested Grade Levels: 4–6

A choropleth map uses colors or shading to show differences between areas and give information about demographics. You can use census data to create choropleth maps. Give students a blank piece of paper and have them randomly draw X's around the paper. Encourage them to cluster X's in certain areas. Next, have the students lightly draw a pencil grid over the paper. Assign specific colors (with lighter colors for smaller numbers) and have students color the grids based on the number of X's in each grid box. For example, a box with 0–2 X's might be colored a light blue, 3–5 X's dark blue, and 6–8 X's purple. Show students a choropleth map of population density and compare to the products they made.

62. Topographic Maps

Materials: colored foam board with sticky back, scissors, enlarged topographic map of an area
Time: 30 minutes
Suggested Grade Levels: K–1; 4–6

A topographic map uses contour lines to show elevation on a flat map. It is a representation of a three-dimensional surface. When contour lines are close together, the terrain is steep. Ask students, "Why would you use a topographic map?" (Possible answers include to choose the best route for a hike, find the best place to build a road, create a plan for natural disasters, plan a subdivision of homes, or conduct energy exploration.) Students can view topographic maps of an area at https://www.topozone.com/.

Younger children can see a good example of a topographic map by using a knuckle and a marker. Put your hand into a fist and draw three parallel lines around the knuckle (I usually put a dot on the top). Open your hand a show the students the flat knuckle rings. Discuss how the "topographic" rings represent the height of the knuckle hill.

Older students can make a topographic map using colored foam board. Show students a topographic map of an area, such as Mammoth Cave. You can enlarge a small section of the map and give a copy to each student, or use the topographic map in appendix H. Have students trace the outline of the edge onto the colored foam board. Cut out the foam shape. Trace the next lines of the map onto the foam board until you reach the smallest section. Peel the back of the foam board and stick together in order to represent the map in 3D.[5]

63. Shipwreck Island[6]

Materials: topographic map of an island, drawing tools, crayons, tea bags, paper towels
Time: 45 minutes
Suggested Grade Levels: 4–6

Tell the students that you have found an old book that tells the story of Black-Eyed Jack, a notorious pirate who was lost with his crew on an island many years ago. Give each student (or group of students) the topographic map of an island in appendix H. Add dots in the area around the edges to represent the beach and a couple of rivers emptying into the water around the island. As you tell the story to the students, have them mark the locations on their maps. (A) The pirates washed up on a beach on the north side of the island. (B) They found a stream of fresh water emptying into the bay. (C) They built a camp on the east bank of the stream on the edge of the beach. (D) They followed the stream to its source, turned west, and built a lookout on the highest hill. (E) They found a hut on the south side of the island, two kilometers in a straight line from the base camp. (F) An old parchment in the hut directed them upstream to a hill fifteen meters above sea level, where they found gold. (G) The lookout spotted a ship on the easterly point of the island. The pirates dragged their gold to the spot, but they landed right in the arms of some British soldiers.

After checking to see if the points are correct, have the students do the following:

- Draw a beautiful compass rose (make sure north is in the correct spot!) and add fearsome sea beasts around the island.
- Tear off bits of papers around the corners and edges of the map.
- Pat the edges of the map with wet tea bags to make the map look antique.
- Dry the map with paper towels.
- Color the map.

64. Clay Landforms

Materials: index cards, play dough or clay, maps that show
physical landscape features
Time: 30 minutes
Suggested Grade Levels: 2–4

Show students maps that display the physical landscape features of a
place. Ask students, "What is the highest mountain in North America?
(Mount Denali in Alaska) What is the longest river in North America?
(Mississippi River) What landforms can you name?" Make index cards
with instructions to create various landforms (e.g., create a *terrace* near
a *spit* that projects into a *bay*; create a *peninsula* in a *harbor* with an
archipelago nearby). Have each student pull a card and then create the
landforms listed out of play dough or clay. Students should be able to
define their landforms as they present their work.

65. Luscious Landforms[7]

Materials: rectangular cake, white icing, red icing, blue icing,
green icing, green coconut, mini chocolate chips, large chocolate chips, Sno-Cap candies, Red Hot candies, index cards
Time: 30 minutes
Suggested Grade Levels: 3–5

Bake a rectangular cake and cover with white icing. Outline the U.S.
shape on top with red icing. Place instructions on index cards for decorating parts of the cake:

- green coconut for prairies
- blue icing for the Great Lakes, Mississippi River, Atlantic Ocean,
  Pacific Ocean, and Gulf of Mexico

- mini chocolate chips for the Ozark Mountains and the coastal ranges
- large chocolate chips for the Appalachian Mountains and Sierra Nevada Range
- Sno-Cap candies for the Rocky Mountains
- Red Hot candies for Washington, D.C., and major cities
- green icing for the Grand Canyon

Students will choose a card and decorate their part of the cake. After all students have finished, eat the cake!

66. Getaway Island[8]

Materials: book *Where the Wild Things Are* (Sendak, 2012), markers, drawing paper
Time: 40 minutes
Suggested Grade Levels: 2–6

Read the book *Where the Wild Things Are*. In the book, Maurice's mother calls him a "wild thing" and banishes him to his room without supper. Maurice travels a night and a day, in and out of a year, to reach an island where he can become king of the wild things. Max uses the island to "get away" from his problems. In this lesson, students create their own getaway island based on a theme, such as Pizza Island. Each island must have at least six geographic features named to represent the theme. For example, Pizza Island may feature the Pepperoni Swamp, the Tomato Sauce River, Thick Crust Volcano, Cheesy Plains, Anchovy Tundra, and Jalapeño Bay. Name the ocean in accordance with the theme. Students need to develop a map key, a compass rose, and a scale, as well as a title for their getaway island. To challenge students, ask them (before reading the book) to tell you all of the landforms they know. List them on the board, then give them the instructions to create their island landforms different from those on the generated list.

67. Street Smarts

Materials: book *And to Think That I Saw It on Mulberry Street* (Seuss, 1989), drawing paper, drawing tools, Google Maps, computer
Time: 45 minutes
Suggested Grade Levels: 3–5

Have students make a list of famous streets, such as Sesame Street, Baker Street (Sherlock Holmes), Privet Drive (Harry Potter), Abbey Road (Beatles), Hollywood Boulevard, Wall Street, and Pennsylvania Avenue. Using Google Maps, investigate famous iconography on the streets. Have students research books, songs, or movies that have street names (e.g., *A Nightmare on Elm Street*, "Penny Lane," *Miracle on 34th Street*, *21 Jump Street*). Ask students to write the names of (a) the street they live on, (b) the street the school is on, and (c) the street of their favorite restaurant. Students can visit the website https://hiconsumption.com/2014/09/29-of-the-worlds-most-famous-streets/ to find twenty-nine of the world's most famous streets. Ask students which streets they might add to the list.

Share the book *And to Think That I Saw It on Mulberry Street*. Have students take one of the streets they listed and draw a map of things they might see on the street. Take a walk on the street around your school. On returning to the classroom, draw fun things they would like to see come down their streets.

68. Mapping Imaginary Places

Materials: books *The Once upon a Time Map Book* (Hennessy and Joyce, 2010) and *Buried Blueprints: Maps and Sketches of Lost Worlds and Mysterious Places* (Lorenz and Schleh, 1999), drawing paper, drawing tools
Time: 45 minutes
Suggested Grade Levels: 3–6

The Once upon a Time Map Book includes maps of Neverland with Peter Pan, Wonderland with Alice, Land of Oz with Dorothy, Aladdin's Kingdom with the Genie, Enchanted Forest with Snow White, and the Giant's Kingdom with Jack. Each map has a key and a series of instructions to find hidden treasures while learning the basics of map reading. A similar book is *Buried Blueprints: Maps and Sketches of Lost Worlds and Mysterious Places*, which includes fourteen detailed drawings accompanied by one-page narratives that offer historical and literary background about places of myth and legend, such as King Arthur's Camelot, Count Dracula's castle, and Atlantis. After sharing the books, teachers can have students create maps of imaginary places from their readings. Be sure to include a title, legend, and compass rose.

69. Mapping a London Epidemic[9]

Materials: map of cholera deaths (https://www.nationalgeographic.org/activity/mapping-london-epidemic/), computer for research
Time: 30 minutes
Suggested Grade Levels: 3–6

Explain to students that scientists and researchers often map disease outbreaks to keep an illness from spreading further. Have students examine a copy of the John Snow maps of cholera deaths and water pumps in London. Ask students, "Where are the most deaths concentrated? Which pump do you think might have had cholera-infected water? Why? What other information might have been helpful for Dr. Snow's analysis? What would be the next steps for someone who used this type of analysis?" Have students do their own research for recent epidemics, such as H1N1 or Ebola.

70. Map Mania

Materials: state cookie cutters, cookie dough, blue icing, chocolate chips, various types of unusual maps, atlases, laminated maps cut into puzzles
Time: 25 minutes for each activity
Suggested Grade Levels: 2–6

Give each student a state cookie cutter to cut out shapes from a roll of cookie dough. Instruct them to use blue icing to mark important lakes and rivers in the state they cut out. Help students place chocolate chips to indicate major cities and towns. Use the school cafeteria to bake the cookies.

Give students unusual maps, such as languages in Russia, the subway in Atlanta, or waterlines in Phoenix. Ask the students to guess the type of map. Instruct students to design a key for the map. Then have students construct their own unusual map.

Have students write the letters of their name down the side of a piece of paper. Then ask students to use an atlas to find a body of water, a city, or a landform that begins with each letter.

Collect maps from old *National Geographic* or road maps and laminate them. Cut the maps into puzzle-shaped pieces and have students put the maps together. Ask students to describe the type of map they put together and how the map could be used.

71. Subway Antics

Materials: Metro subway map (https://www.wmata.com/schedules/maps/upload/2019-System-Map.pdf)
Time: 30 minutes
Suggested Grade Levels: 4–6

Have the students pretend they just arrived at the airport in Washington, D.C., and they must take the Metro to get to their hotel at McPherson Square. Encourage students to plan two different routes, in case one of the subway lines is shut down for repairs. From the hotel, what is the best way to get to Arlington Cemetery? the Pentagon? the Smithsonian?

72. What Constitutes a Good Map?[10]

Materials: various types of maps (road, political, relief, satellite, pictorial, historical)
Time: 30 minutes
Suggested Grade Levels: 4–6

Divide students into groups and give each group a series of maps. Ask students to rank the maps from best to worst. Students must come up with a definition as to what represents a "good" map and invent categories with which to do the ranking. Ask each group to explain their rankings of the maps. Discuss the different categories created by the students. Students may infer that maps are judged by use, and that culture may affect the judgment.

ANNOTATED CHILDREN'S LITERATURE FOR TEACHING TYPES OF MAPS

Hennessy, B. G., and Peter Joyce. *The Once upon a Time Map Book*. Cambridge, MA: Candlewick Press, 2010.
  This book includes maps of Neverland with Peter Pan, Wonderland with Alice, Land of Oz with Dorothy, Aladdin's Kingdom with the Genie, Enchanted Forest with Snow White, and the Giant's Kingdom with Jack. There is a key and a series of instructions so that young readers can follow paths to hidden treasures while learning the basics of map reading.
Knowlton, Jack. *Maps & Globes*. New York: Collins Publishers, 1986.
  This book chronicles the history of mapmaking, from scratches in the sand to clay maps and stick chart maps. This book also includes simple

explanations of how to read maps and globes and introduces many different kinds of maps.

Lorenz, Albert, and Joy Schleh. *Buried Blueprints: Maps and Sketches of Lost Worlds and Mysterious Places.* New York: Harry N. Abrams, Inc., 1999.

This book has fourteen detailed drawings accompanied by one-page narratives that offer historical and literary background about places of myth, legend, and literature. The elaborate sketches portray such places as the Seven Cities of Gold, King Arthur's Camelot, Count Dracula's castle, the Garden of Eden, and Atlantis. The premise is that the book is based on the travels of Albert Lorenz, and even includes a magnifying glass to use while exploring the intricate details of each page.

Rey, H. A. *Curious George Lost and Found.* Boston: HMH Books for Young Readers, 2008.

Curious George accidently gets lost on a river and must find his way back home by creating a map of landmarks. Not only does he save himself, but he also helps others get home. The book shares ideas on building a raft and creating a treasure map.

Ritchie, Scot. *Follow That Map! A First Book of Mapping Skills.* Boston: Kids Can Press, 2009.

Sally and her friends take a trip through their neighborhood, city, country, and world to find a missing cat and dog. The book explains key mapping concepts and includes an activity of how to make a map of your bedroom.

Sendak, Maurice. *Where the Wild Things Are.* New York: HarperCollins Publishers, 2012.

Maurice's mother calls him a "wild thing" and banishes him to his room without supper. Maurice travels a night and a day, in and out of a year, to reach an island where he can become king of the wild things.

Seuss, Dr. *And to Think That I Saw It on Mulberry Street.* New York: Random House Books for Young Readers, 1989.

The notable rhymes and illustrations of Dr. Seuss shine in this book as a young boy imagines the interesting characters that go by on Mulberry Street. Marco uses his imagination to find interesting sights on his way to and from school.

Taylor, Barbara. *Maps and Mapping.* New York: Kingfisher Books, 1995.

This book is more content specific with explanations of how to use, draw, and understand maps. Concepts include scales, symbols, contour lines, angles, latitude and longitude, map projections, compass directions, aerial photographs, grids, and satellite photographs.

Wade, Mary Dodson. *Types of Maps* (Rookie Read-about Geography: Maps and Globes). Chicago: Children's Press, 2003.

Various types of maps and their uses are presented in this easy-read series. Young readers are introduced to road maps, city maps, political maps, elevation maps, product maps, historical maps, and map keys.

ASSESSMENT QUESTIONS—TYPES OF MAPS

- What are some different types of maps?
  Possible Answer: topographic, thematic, geological, transit, physical, political, weather
- Why do we have different types of maps?
  Possible Answer: Maps have different uses. A general reference map would help you get to a specific destination. A topographic map would show elevation. A thematic map would identify information on specific topics, such as rainfall in an area.
- Who can use maps?
  Possible Answer: Anyone can learn to use a map. Maps are useful to find out important information, search for directions, measure distance, search for campgrounds, show settlement patterns, and give input on the weather.
- How do landforms affect us?
  Possible Answer: Landforms affect climate and affect where humans choose to live. They affect what activities we can do.
- How would a topographic map differ from a street map? a satellite map?
  Possible Answer: Topographic maps show elevation. Street maps would show roads and destinations. Satellite maps are images taken from satellites.
- Compare the Robinson and Mercator map projections. Which would be more accurate?
  Possible Answer: The Robinson image has a better balance of size and shape of high latitude lands. With the Mercator map, Greenland and Africa are portrayed as about the same size, whereas in reality Africa is almost fourteen times larger.
- When would a cartogram be appropriate to use? a choropleth map?
  Possible Answer: A cartogram combines statistical information with geographic location and provides emphasis on certain infor-

mation. A choropleth map also measures statistics; however, the areas are color coded rather than exaggerated.

- What types of maps can you find in our school?
  Possible Answer: In my school, you can find the Mercator and Robinson map projections.
- Who might use a satellite map?
  Possible Answer: Meteorologists use satellite images to forecast the weather. Foresters use images to monitor fires. Urban and rural planners use images for property development.
- Why should a story include a map?
  Possible Answer: A map is a guide to all the places a story takes you. It helps the writer craft the story.
- What is a thematic map?
  Possible Answer: Thematic maps are developed to show a unique topic, such as religions of the world.

NOTES

1. "History of Mapping," Intergovernmental Committee on Surveying and Mapping (ICSM), accessed November 4, 2019, https://www.icsm.gov.au/education/fundamentals-mapping/history-mapping.

2. Adapted from lesson idea by Christopher Stanton, teacher consultant with the Vermont Geographic Alliance.

3. Adapted from lesson "Where There's Light, There Are People," in *Teacher's Handbook for Geography Awareness Week* (Washington, D.C.: National Geographic Society, 1998).

4. Adapted from lesson "World Population Map: A Cartogram for the Classroom," Population Education, accessed November 4, 2019, https://populationeducation.org/world-population-map-cartogram-classroom/.

5. Adapted from lesson idea by Kathleen Matthew, professor, Western Kentucky University.

6. Adapted from lesson "Shipwreck Island," *Geography Assessment Framework for the 1994 National Assessment of Educational Progress*, NAEP Geography Consensus Project, Contract Number RN91073001, Council of Chief State School Officers, National Council for Geographic Education, National Council for the Social Studies, American Institutes for Research, 1994.

7. Adapted from lesson idea by Sandra Goldich, teacher consultant with the Louisiana Geography Education Alliance (LaGEA).

8. Adapted from lesson idea by Shelley Williams, teacher consultant with the Louisiana Geography Education Alliance (LaGEA).

9. Adapted from lesson "Mapping a London Epidemic." National Geographic Education, accessed November 4, 2019, https://www.nationalgeographic.org/activity/mapping-london-epidemic/.

10. Adapted from lesson "What Constitutes a Map?" Avner Segall, "Maps as Stories about the World," *Social Studies and the Young Learner* 16, no. 1 (2003): 21–25.

## CHAPTER 7

Mapping with Technology

Figure 7.1. *Maps and Technology*
Madalyn Stack, artist

The role of technology has grown over the years and can be used as a powerful tool for geography instruction. Satellites in space can send back images of the earth in minute detail. This is very helpful to cartographers, especially for mapping places difficult to reach, such as mountains or rain forests, or tracking the weather. Satellite monitoring of the environment has been invaluable to concerned citizens who wish to preserve unique ecosystems. Sonar (sound navigation and ranging) technology is used to map the ocean floor. Many government agencies now allow stakeholders to use online mapping tools to map census data, examine revitalization efforts, plan community events, or monitor environmental cleanup.[1]

Geographic Information System—GIS

As the role of technology in the classroom has increased over the years, teachers are looking for ways to integrate tools in meaningful ways. Geographic Information System (GIS) technology allows geographic information to be stored, analyzed, and manipulated. Students can take actual data from local surroundings and create GIS projects to solve problems in their own environment. GIS combines layers of information to give a better understanding of place. Sample layers you can put on a map include streets, parks, hospitals, banks, rivers, airports, restaurants, railroads, crime scenes, bridges, or landforms. Types of data for GIS can include base maps (streets and highways, rivers and lakes, parks and landmarks, place names, boundaries); business maps and data (census/demography, consumer products, financial services, health care, real estate, crime, transportation, telecommunications); environmental maps and data (weather, environmental risk, satellite imagery, topography, natural resources); and general reference maps (world and country maps). This interactive mapping tool can be used to identify patterns, linkages, or trends and give a better understanding of place.

GIS is used for real-world projects, such as studying the spread of infectious diseases, tracking wolves, managing mock evacuation of citizens, identifying traffic patterns, planning school bus routes, mapping land use to determine the need for green space, studying the effects of global warming, and tracking weather fronts and hurricanes. In a GIS project for the classroom, students would use the inquiry approach to formulate research questions, develop methodologies, gather and analyze data, and draw conclusions. Begin by framing a question: Where were most of the burglaries last month? Select data from government organizations, commercial data providers, or the Internet. Choose an analysis method, such as mapping individual crimes and crime areas over a specific time period. Students would process the data by looking at location, quantity, and area measurement. Finally, students would present the results to stakeholders and determine an action to be taken.

Every day, choices tied to locations are made in communities across the country. GIS technology can help students identify specific data items that can be tied to their everyday world, and display the geographic data in a variety of ways. Esri (Environmental Systems Research Institute) provided a grant in 2014 to make mapping software available free to elementary, middle, and high schools in the United States, in response to President Barack Obama's call to help strengthen STEM education.[2] With the publication of GIS books for elementary students, now even first graders can solve spatial problems when introduced to GIS technology.

Global Positioning System—GPS

A Global Positioning System (GPS) is a navigation and positioning tool that connects to at least four (of the twenty-four) satellites to determine your exact location (latitude/longitude). GPS can be used to show how far you have hiked or track your way back home. It can help you find the nearest airport or hospital. An outdoors person might use GPS for exploring, sailing, flying, or biking. Hikers, hunters, and anglers often use GPS systems to find destinations, mark the journeys they have traveled, or just note specific locations.[3] GPS can be used to help settle property disputes between landowners. The U.S. Department of Defense first put GPS satellites into orbit for military use, but by the 1980s civilians were allowed to use them.

Geocaching and EarthCaching

Geocaching is an outdoor "treasure hunt" activity in which the participants use a GPS system to locate containers (called "geocaches") anywhere in the world. The very first geocache was hidden by a computer consultant in Oregon, who posted the coordinates online and challenged others to find it. Geocaches are currently placed in countries all over the world and on all seven continents, including Antarctica. In my city alone, there are more than three hundred caches. Often a small prize is placed within the cache. A general rule is to leave something for the cache if you take something. The finder should document online at www.geocaching.org (or in a notebook in the cache) that the cache was located.

Similar to geocaching, EarthCaching involves finding then studying and compiling information about landmarks or locations. Students use GPS technology to find sites through www.earthcache.org and follow coordinates to learn and appreciate something about the earth. For example, at the South Cumberland State Park in Kentucky, students will find a natural rock bridge. At the website, students are to post a photo of themselves under the bridge and answer the following questions: How thin is the thinnest point on the bridge? Where does the spring currently flow? How long do you think it took to create this bridge? This is a great way to get students outdoors to explore nature.

Remote Sensing

Many technologies allow us to gather information about a place at a distance, and allow us to see farther and observe finer details. Aerial photographs were first taken from hot air balloons in the 1860s. Infrared film and radar came into use during World War II. The first satellites were launched in the 1950s and 1960s, initially focused on providing images of clouds. By stepping into space with remote sensing, students can view the ecosystems, land patterns, physical processes, and other features of their home at a global scale.

Activities for mapping with technology can be tied to literature, communities, and sandwiches. Students in this day and age are very tech savvy and will enjoy using these tools to find out more about the world in which they live.

ACTIVITIES FOR TEACHING MAPPING WITH TECHNOLOGY

73. Geocaching

Materials: Internet, phone or camera to take photos, transportation to site
Time: 1–3 hours
Suggested Grade Levels: 4–6

Encourage students to find a geocache in their area. Put in the zip code on the geocaching website (www.geocaching.org) to get latitude and longitude markings for caches. Have students take photos of themselves at the caches and describe the locations. Even better is to have students create their own caches. Discuss with students which special place they think would make a good location for a cache and why.

74. EarthCaching

Materials: Internet, transportation to site
Time: 1–3 hours
Suggested Grade Levels: 4–6

Encourage students to find an EarthCache in their area (www.earth-cache.org). The Geological Society of America runs the EarthCaching site and has developed a series of guidelines for anyone posting on the site. Caches must provide an earth science lesson and be educational. There must be landowner or land manager permission for these unique locations. Have students decide other special places from their area that might make a good EarthCache place.

75. GIS Sandwich[4]

Materials: various colors of construction paper, scissors
Time: 45 minutes
Suggested Grade Levels: 3–6

Ask students, "What can you do with GIS?" (You can plan a school bus route, decide where to build a new coffee shop, track wolves across Yellowstone Park, study spread of a disease, find your way to a pizza place, monitor drug arrests near schools.) Have students cut out construction paper to represent parts of a sandwich (lettuce, tomato, pickle, cheese, bread). Give students the problem: Where is the best place to build a new coffee shop in town? What data could you use to solve the problem? Use the parts of the sandwich to represent each layer of data that you might choose on the GIS program (e.g., lettuce = location of other video stores; tomato = empty buildings or lots; cheese = socio-economic data; pickles = traffic flow). Let each student describe the layers of their sandwiches.

76. GeoScavenger Hunt

Materials: camera, map of town, GPS app
Time: 1 day or weekend
Suggested Grade Levels: 3–6

Take photos of places around town. Insert the photos into a word document and give to students, along with a map of the town. Tell the students that they must each find at least three of the places pictured, mark the latitude/longitude of each place (with a GPS or phone app), mark the place on the map, and take a photo of themselves at each place. This can be a weekend project. You could have your students create their own photo scavenger hunt around the school or community.

77. Earth from Space

Materials: book *Seeing Earth from Space* (Lauber, 1990), Internet, computers
Time: 30 minutes
Suggested Grade Levels: 3–6

Read to students *Seeing Earth from Space*, which features real photographs of Earth from space. Show students the USGS website images of the earth collected by remote sensing (http://remotesensing. usgs.gov/gallery/). By using infrared sensing, cartographers can detect temperatures of objects on the ground. This can help with mapping living objects, such as crops. Have students examine satellite maps of the oceans and identify which are warm areas and which are cool areas. Students are often interested in examining remote sensing images of hurricanes, volcano eruptions, or other natural disasters.

78. Batty Images

Materials: book *The Adventure of Echo the Bat* (Butcher and Broadhurst, 2001), Blue Marble image of Earth (https:// www.nasa.gov/content/blue-marble-image-of-the-earth-from-apollo-17), computers
Time: 30 minutes
Suggested Grade Levels: 2–4

Read to students *The Adventure of Echo the Bat*, which incorporates Landsat satellite images into the story of a bat migrating through habitats of Arizona (an online version can be found at http://science. hq.nasa.gov/kids/imagers/intro/story.html). The satellite imagery encourages students to examine perspective, pattern, color, texture, and shape. Have students examine the Blue Marble image of Earth from space taken by Apollo 17 astronauts in 1972. Students can collaborate to design presentations about the earth's cycles and how humans have impacted the planet. Show students satellite images of their own communities by using https://www.google.com/earth/.

79. Story Mapping

Materials: Internet, computers
Time: several days
Suggested Grade Levels: 4–6

Have students research various aspects of their community, such as monuments and memorials, festivals held in the area, government buildings, or economic enterprises. Once students decide on a topic, have them develop an online, interactive, annotated map of their community with text and photographs. Information on how to make a story map can be found at https://storymap.knightlab.com/ or https://story-maps.arcgis.com/en/how-to/.

80. My Home, School, and Community[5]

Materials: book *The Little House* (Burton, 1978), GIS program, computers
Time: 1 hour
Suggested Grade Levels: 4–6

Read aloud to students the book *The Little House*. This book tells the story of how a little house once located in the country became the center of a city through urbanization. Have students use GIS to examine their own city and answer questions such as these: Was your school in the city limits in the 1800s? Is there any difference in the size of your city between the 1800s and today? How have the buildings and roads changed over time? In which direction has your city grown the most? What seems to influence the growth patterns?

81. MapQuest

Materials: Internet, computers
Time: 45 minutes
Suggested Grade Levels: 3–6

Have students plan a journey to Disney World using a mapping program, such as maps.google.com or mapquest.com. Find at least two different ways to travel. Incorporate travel by bus, car, airplane, and walking. Ask students, "How many hours will it take to arrive? How many miles will you travel? How will the cost vary between car, bus, and airplane travel? Would it be possible to travel the distance by train?"

82. The Amazing Race

Materials: GPS device; large park or playground area; jump rope, Hula-Hoops, balls, cones, or other materials to create a movement task
Time: 2 hours
Suggested Grade Levels: 4–6

Determine several locations around an area, such as a park, football field, or playground. Divide the students into teams and have them race to various locations using their GPS devices. At each location, provide a task for each team to complete (e.g., jump a rope ten times, shoot a ball through a hoop, run around a cone, or do a somersault). The first team to complete the tasks successfully is the winner.

83. Scavenger Hunt from Space

Materials: Internet, computers
Time: 1 hour
Suggested Grade Levels: 3–6

Divide students into groups. Give each group the latitude/longitude coordinates of a different building in their community. Have students type the coordinates into the Google Earth app. As students zoom in and locate the "absolute" location of their building, they need to develop three clues for the building based on "relative" location. Each group can share their clues and allow the other groups to guess the name of the building. Use the chart in appendix I to set up the lesson.

84. Community Life[6]

Divide students into small groups to research their own community. Tell students that they are to find the best sites for three different families with varying requirements to live. The sites should be predetermined to include a variety of housing and to allow students to give persuasive arguments for their decisions. Using GIS, students should look at school locations, house types, amenities, traffic, nearby services, noise levels, crime rates, and so on. Ask students questions, such as, "Why do you think people would want to live at this site? What factors helped you to make your decision?"

As a follow-up activity, allow students to create a promotional video about the area. Have them consider what makes their community special and would look good on film. Encourage students to think about what there is to do in the community, why people might like to live there, what transportation networks are available, and what things are nearby.

85. Learning How to Pan[7]

Have students make a "map viewing frame" from four wooden popsicle sticks taped together to form a picture frame. Place a large map on the floor and ask students to slide their frames over the surface of the map, much like a camera pans the landscape. Go to www.mapquest.

com and enter the address of the school. Click the "Get Map" button to see the map. Move the cursor over the map. It changes from an arrow to a hand icon known as the "pan" tool. Have students use the tool to shift the frame of the map. Place the pan tool at any spot on the map and click. That point becomes the new center of the map.

86. Learning How to Zoom[8]

Materials: handheld magnifying lenses, large map, Internet
Time: 25 minutes
Suggested Grade Levels: 1–3

Allow students to observe a large floor map with handheld magnifying lenses. Then go to www.mapquest.com and enter the address of the school. Along the right side of the map is the zoom tool. Click a lower step to zoom out. Ask students, "What new information becomes available as you zoom out? What labels disappear? What happens to the scale bar?"

ANNOTATED CHILDREN'S LITERATURE FOR TEACHING MAPPING WITH TECHNOLOGY

Bramley, Paul. *Mapping Skills with Google Earth: Grades PK–2*. San Diego, CA: Classroom Complete Press, 2011.
This book provides activities for students in Pre-K through grade 2 with mapping skills. Handouts, rubrics, and quizzes are included.

Burton, Virginia Lee. *The Little House*. Boston: HMH Books for Young Readers, 1978.
This is the story of a little house that first stood in the country, but the city grew closer and closer until the little house disappeared into the chaos. At the end, someone buys the house and moves it out to the country.

Butcher, Ginger, and Beth Broadhurst. *The Adventure of Echo the Bat*. Washington, D.C.: U.S. Government Printing Office, 2001.
Echo's mother teaches him how to use his senses, which he has to use when he gets lost in a storm. The book uses satellite images for exploring each habitat Echo visits on his journey to his winter hibernation place.

Cheshire, J., and O. Uberti. *Where the Animals Go: Tracking Wildlife with Technology in 50 Maps and Graphics.* New York: W. W. Norton & Company, 2017.
This book uses data from satellites, drones, cameras, and other technology to document how creatures like otters, owls, turtles, and sharks navigate the world. The beautiful illustrations document the role new technologies play in expanding our knowledge of animal wanderings.

Lauber, Patricia. *Seeing Earth from Space.* New York: Orchard Books, 1990.
This book is a compilation of photographs taken by astronauts and satellite images of Earth. The author describes ways in which scientists can use remote sensing to analyze evidence of damage to the planet.

ASSESSMENT QUESTIONS—MAPPING WITH TECHNOLOGY

- What is the difference between GIS and GPS?
  Possible Answer: GPS uses satellites to receive information to help people determine their location. GIS is a software program that allows people to place map layers and visually represent data.
- Who would use GIS or GPS and why?
  Possible Answer: Everyone can use GIS or GPS to play games (like Pokémon Go), hike, pilot an airplane, map for an emergency, locate crime areas, predict the outbreak of disease, or note traffic patterns.
- If you could make a geocache, where would it be and why?
  Possible Answer: I would make a geocache at Lost River Cave in my community. It is a great family vacation center and visitors can ride a boat into a cave.
- How can using maps and technology benefit your school? the community?
  Possible Answer: Technology is already a big part of our culture. The use of GIS and GPS prepares us for future careers, engages students, allows access to the most up-to-date information, encourages problem-solving, and initiates decision-making skills. In the community, this technology can identify at-risk or underserved populations, help with management of natural resources, demonstrate high-risk areas, and allow for better city planning.

- Who would use a story map?

  Possible Answer: Environmentalists can use story maps to learn about world treasures, activists can use story maps to create change in society, business people can use story maps to create a compelling presentation, and teachers can use story maps for instruction.

- What important places would you put on a story map for your community? Why?

  Possible Answer: For my community, I will include the tourist sites on my story map, the important businesses in the area, the university, natural features, and historical sites nearby.

- How can a Google Map help you plan a vacation?

  Possible Answer: A Google Map will allow you to create a custom map and add pins for locations you want to visit. You can get websites and other information about the places you want to visit. You can add notes and photos on your map as you visit places. You could include bike path routes and add icons to mark places you specifically want to visit. You can also share your map with others and access it from your phone.

- How can satellite images be used?

  Possible Answer: Satellites are "eyes in the sky" and can be used to get images. These images can be used to understand landscape change, monitor ocean temperatures, inventory effects of landslides, monitor earthquakes, note population growth, gather intelligence, and monitor nuclear power plants.

- What is the purpose of remote sensing?

  Possible Answer: Remote sensing can monitor large forest fires, track clouds to predict weather, monitor volcanos, map the ocean bottom, and track changes in farmlands.

- How is a magnifying glass similar to the "zoom" function?

  Possible Answer: A magnifying glass produces a magnified image of an object. The zoom function gets closer on a map to make it bigger in a certain area.

- How is a picture frame similar to the "pan" function?

  Possible Answer: A frame highlights a certain area of an object. The pan function moves the map to the area you want to focus on.

NOTES

1. Sarah Witham Bednarz, Gillian Acheson, and Robert S. Bednarz. "Maps and Map Learning in Social Studies," *Social Education* 70, no. 7 (2006): 398–404, 432.

2. "Esri Pledges $1B in Cloud-Based STEM Software to White House ConnectED Initiative," *The Esri Newsroom*, May 27, 2014, accessed November 4, 2019, http://www.esri.com/esri-news/releases/14-2qtr/esri-pledges-1b-in-cloudbased-stem-software-to-white-house-connected-initiative.

3. Bednarz et al., "Maps and Map Learning."

4. Adapted from lesson "Making an Earth Sandwich," *Explore Your World with a Geographic Information System: A Teaching Supplement for Grades 5–12 Introducing Basic GIS Concepts and Components.* Environmental Systems Research Institute, Inc. (1995): 9, accessed November 4, 2019, https://www.scribd.com/document/22344623/explore-your-world-with-a-GIS-teachers.

5. Adapted from lesson idea by Eui-kyung Shin, "Using Geographic Information System (GIS) Technology to Enhance Elementary Students' Geographic Understanding," *Theory & Research in Social Education* 35, no. 2 (2007): 231–55, DOI: 10.1080/00933104.2007.10473334.

6. Adapted from lesson "Location, Location, Location," by Elysia Parnell, "Geography Is All about Location, Location, Location! "*Teaching Geography* 32, no. 2 (2007): 91–92.

7. Adapted from lesson "Panning" by Marsha Alibrandi, "How to Do It: Online Interactive Mapping: Basic Activities," *Social Studies and the Young Learner* 17, no. 3 (2005): P1–P8.

8. Adapted from lesson "Zooming," Alibrandi, "Online Interactive Mapping."

## CHAPTER 8

Mapping with the Five Themes of Geography

Figure 8.1. *Five Themes-Movement*
Madalyn Stack, artist

In 1984, the *Guidelines for Geographic Education*[1] introduced big ideas in geography known as the Five Fundamental Themes of Geography. These five areas of investigation are appropriate for teaching geography in the early grades. The themes are an integrative framework for studying the world:

- *Location*: where a point is on the earth's surface and why it is there

- *Place*: special features an area might have, such as climate, people, and landforms
- *Human/Environment Interaction*: how people react to and sometimes change their environment
- *Movement*: travel and communication with one another; movement of products and ideas
- *Regions*: areas of the earth that are alike in some way or another

Elementary teachers can remember the acronym MR. HELP for the five themes of geography (*M*ovement, *R*egion, *H*uman/*E*nvironment interaction, *L*ocation, *P*lace).

With location, students can identify precise points on the earth by using the grid system of latitude and longitude. This is known as absolute location. Street names, house numbers, and school room numbers are also examples of absolute location. Relative location would refer to the position of a place in relation to other places. Students can relate that their houses are near the school, around the corner, or at the center of town. Typical questions for location include, "Where is this place? What is it near? How do I get there from here? Why is this place located here?"

Place is described by physical (climate, soil, water, topography) or human (language, religion, population distribution) characteristics. Place can often be described by the culture of people who live there, although there are some places untouched by human contact. Each location is unique based on the physical and human characteristics that exist there. Typical questions for place include, "What qualities make a place desirable? What is this place like? What physical and human features does it have?"

Human/Environment Interaction addresses the ways that people continually modify or respond to the environment. These interactions often have intentional and unintentional consequences. Strip mining, land use changes, and slash-and-burn practices can serve as springboards for discussions on responsible behavior and attitudes toward environment interactions. Typical questions for human/environment interactions include, "How are people's lives shaped by the place? How has the place been shaped by people?"

Movement relates to the global interdependence of today's society. As people travel and communicate further and faster, students need to realize our mutual dependence on one another. Products and information travel across the globe, and there are costs and benefits related to this movement. Trade can often be restricted, and free trade can be a controversial topic. Typical questions for movement can include, "How does international trade influence the lives of American citizens? How did people, products, and ideas get from one place to another? Why do they make these movements?"

Regions are a context for defining and studying places by examining common features. Characteristics such as language, landforms, ethnicity, religion, and types of government could be used to define a region. Smaller regions can lie within larger ones. The United States is divided into regions according to their geographic location: the Northeast, Southwest, Southeast, Midwest, and West. Typical questions for region can include, "How can we characterize the South? How is the South different from the Northeast? What defines the Rust Belt? What features set this place apart from other places?"

It is quite simple to integrate the five themes into examples easily understood by students. The elementary school can be found at an exact location on a map. A school has describable characteristics that indicate its place. There is an area from which the school draws its population (region). The faculty and students at a school interact with one another and the landscape. The area within and around the school is constantly changing, which involves human/environment interactions. There is a constant movement of people and goods in and out of the school.[2] Using the five themes helps students understand the connections between people and places in the world.

ACTIVITIES FOR TEACHING MAPPING WITH THE FIVE THEMES OF GEOGRAPHY

87. Where to Build?[3]

Materials: map of an area
Time: 20 minutes
Suggested Grade Levels: 4–6

Where might be a good place to build a house? Why? Give students a map of an area and ask them to describe characteristics of the place that might indicate a good place to build a factory. Have students describe in terms of place, location, region, human/environment, and movement.

88. Guessing Locations[4]

Materials: none
Time: 20 minutes
Suggested Grade Levels: 4–6

Have students list three physical and human characteristics (e.g., island country, Big Ben, River Thames) to describe five well-known locations (e.g., London; Washington, D.C.; San Francisco; Paris; Chicago). By teams, have the students guess the place in the fewest number of clues.

89. Positive/Negative[5]

Materials: community maps
Time: 25 minutes
Suggested Grade Levels: 4–6

Ask students to consider these questions: "What if a shopping mall was eliminated from or added to our community? How would this change the environment? What are the positive and negative effects of building factories, malls, dams, or sports centers on a community?" Give students copies of their community maps. Ask students to determine the best place to build a new Starbucks. Invite students to justify their decisions. Make connections to location, place, movement, and human/environment interactions.

90. Space Station on the Moon[6]

Materials: drawing paper, markers
Time: 30 minutes
Suggested Grade Levels: 3–5

Ask students to consider the following questions: How do the presence of railroads and airports affect the movement of people and goods in and out of this region? How do satellite dishes and radio towers add to movement of ideas? What other forms of movement are found in our community? How would our lives be different without one of these forms?

Encourage students to design a map of a space station on the moon. Include physical and human features. Divide the station into regions. Introduce communication protocols for the community. Explain how goods could be moved throughout the community.

91. What Is a Region?[7]

Materials: drawing paper, markers
Time: 30 minutes
Suggested Grade Levels: 2–4

Define the term "regions" for students (an area having definable characteristics). Introduce various regions in the United States: Badlands, Everglades, Great Plains, and Death Valley. Ask students to tell you what characteristics define these regions. Remind students that communities have residential, industrial, and commercial regions. Tell students that the classroom can also be divided into regions. Have students brainstorm various regions of the classroom (e.g., art center, reading center, teacher center). Give students art paper and ask them to draw and label regions of the classroom.

92. Build Your House

Materials: book *How a House Is Built* (Gibbons, 1990), book *The Little House* (Burton, 1978), pictures of various types of houses, drawing paper, markers
Time: 25 minutes
Suggested Grade Levels: 2–4

Collect pictures of all types of houses around the world to show to students. Ask students to describe the types of houses they live in. Read the book *How a House Is Built*, then ask students to draw the floor plans of their own houses. Have students draw routes for emergency meeting places outside in case there is a fire. Discuss how the environment might influence the type of building materials used for a house. Share *The Little House* to show how a house moves from a rural to an urban area through the building of a city around it.

93. Baseball and the Five Themes[8]

Materials: U.S. maps for each group of students, Google Maps to find latitude/longitude
Time: 45 minutes
Suggested Grade Levels: 5–6

The sport of baseball is a great way to use mapping with the five themes of geography. Ask students to map the location of pro baseball teams. Then find the latitude and longitude of each city. Find the street addresses of the baseball stadiums. Students might consider why the stadiums and teams are located at these places.

Next, ask students to consider place. What does the land around the stadiums look like? What natural features are found in the area of the stadiums? What is the population of each city? With human/environment interaction, ask the students to consider these questions: How did climate affect the design of the stadiums? How have the stadiums affected the cities in terms of transportation?

The last two themes relate to movement and regions. Ask the students: "What transportation sites are available to allow people to travel to the stadiums? How do the teams travel to other cities to play games? Where does the food sold in the stadiums come from? Where do other teams come from to play in the stadium? In what region might Dodger fans be found? Ranger fans? Pirate fans? Why do some regions have more teams than other regions? What are some regions for teams?"

Students might also enjoy identifying mascots for the professional sports teams. You could have students select places studied in social studies, draw a mascot to represent the place, and explain why the mascot is suitable. Students could even write a slogan or pep song that incorporates the mascot and geographic features of the places.[9]

94. Five Themes Book Map

Materials: book that student has read, drawing paper, markers
Time: 30 minutes
Suggested Grade Levels: 3–6

Encourage students to create a map book report of a book they have recently read. After drawing a map that represents a place in the story, have the students mark a specific location, describe what things look like there (place), indicate how humans changed the environment or the environment changed humans, show something moving and describe where it is going, and identify characteristics that define this place (regions). Students can then hold up their map as they make a five themes book report.

95. My Special Place

Materials: drawing paper, drawing tools
Time: 30 minutes
Suggested Grade Levels: 2–4

Encourage students to think of a place that is special to them. Have students imagine the environment of the place; the location of the place; any specific people, plants, or animals that might exist in that place; and any prominent features of the place. Using the five themes of geography, have the students map their special place. What makes it special? Would you want others to go to your place? Why or why not?

96. Playground Project[10]

> **Materials:** camera, sketch pads, pencils, drawing paper, markers
> **Time:** 1 afternoon
> **Suggested Grade Levels:** 4–6

Introduce a project to students to make their school playground more welcoming and a more useful environment for all. Take students outside to the playground and have them sketch equipment, areas for specific activities, and boundaries that are already present. Some students can take photographs from different vantage points. Back in the classroom, have students compare the photographs and sketches and discuss what they like about the playground. Discuss with students ways to improve the playground so that students with different physical abilities, skills, or cultural backgrounds can use and enjoy the equipment and areas. Introduce ideas for creating various regions on the playground, such as a garden area, music area, building area, water table, or gymnastics area. Give students art paper and have them design their new playground. Have students vote on what they like best about each project.

97. 3D Playground[11]

> **Materials:** camera, cardboard, clay, paint, buttons, spools, rocks, popsicle sticks; flour, salt, water, cream of tartar, food coloring (salt dough map); salt, water, flour, liquid soap, paint (baker's clay)
> **Time:** 1 afternoon
> **Suggested Grade Levels:** K–2

Take students outside to visit the playground. Take pictures of the playground and talk about the equipment and play areas with the students. Make connections to the five themes of geography. Back in the classroom, have students build a 3D map of the playground using all of the art materials. You may want to take the students back out to the playground several times to take a second look and refresh their memories.

If you want to try a salt dough map, mix together 4 cups flour, 2 cups salt, 2 cups water, 2 tablespoons cream of tartar, and food coloring. Students can mix together with their hands and form the shapes. Once dry, the dough can be painted if you don't want to use food coloring.

If you want to try baker's clay, mix together 1 cup salt, 1½ cups warm water, 4 cups flour, and a few drops of liquid soap. Students can create sculptures to bake (325 degrees for 30 to 45 minutes). The clay sculptures can be painted when cool.

98. Five Themes Photographs[12]

Materials: photographs of a town, Google Map of the town, variety of travel photographs from magazines
Time: 30 minutes for each activity
Suggested Grade Levels: 3–5

A good way to allow students to experience other places is to use photographs of these places in the classroom. After introducing students to the five themes of geography, divide the students into groups and distribute a set of photographs to each group. Explain to students that they are to answer questions based only on what they can see in the pictures. Ask questions such as these: "What kind of physical geography do you see? What kind of human geography do you see? How have humans affected the environment in this town? How has the environment affected humans in this town? What jobs do people who live here have? How do they depend on the physical environment?" Pull up a map of the town from the Internet. Ask the students to look at the physical attributes of the town from the map. You can show a satellite image with Google Maps. Ask students to make generalizations about this town.

Give students photographs from travel magazines and have them consider how each of the five themes is represented in the photograph. For example, New York has an absolute and relative location, the water and buildings in New York show place, docks show human/environment interaction, boats show movement, and an aerial view can highlight the harbor as a region.[13]

99. The Three Bears[14]

Materials: cutouts of 3 bears, Goldilocks, 3 beds, 3 chairs, table, 3 bowls of porridge for each child, map of three bears' house, book *Goldilocks and the Three Bears* (Brett, 1996)
Time: 30 minutes
Suggested Grade Levels: K–1

Read the story and then have students act out the story of the three bears, emphasizing the location of the chairs, beds, and table. Give each student a map and the cutouts and allow them to glue the props on the house map. Students can then move the bears and Goldilocks as they retell the story. Discuss each "region" of the house (e.g., bedroom area, kitchen area, living area). Ask students what it might be like if their houses were in a place like the woods.

100. Find the Gingerbread Man[15]

Materials: book *The Gingerbread Man* (Yerrill, 2018), boxes of Little Debbie Gingerbread Cookies (or ingredients to make gingerbread men), school map
Time: 1 hour
Suggested Grade Levels: 2–4

Read the story of the gingerbread man to students. Talk about the five themes of geography as you read the story. Where was the gingerbread made? How did he travel? What did the places look like on his travels? Tell the students that the gingerbread man is hiding somewhere in the school. They must use the school map and go to places marked on the map to try to find the gingerbread man. Have someone return the gingerbread man to the classroom while students are searching. When students come back, show the cookie, and allow all students to eat a gingerbread man.

ANNOTATED CHILDREN'S LITERATURE FOR TEACHING MAPPING WITH THE FIVE THEMES OF GEOGRAPHY

Brett, Jan. *Goldilocks and the Three Bears*. New York: Putnam & Grosset, 1996.

This familiar tale is beautifully illustrated and full of details. The bears are not Papa, Mama, and Baby, but rather, Great Huge Bear, Medium-Sized Bear, and Small Little Wee Bear.

Burton, Virginia Lee. *The Little House*. Boston, MA: HMH Books for Young Readers, 1978.

The little house sits on a hill in the country until the city began encroaching into the area. First a road appears, then carriages, trucks, and other transportation. Soon, apartment buildings and garages surround the little house. One day, someone buys the house and moves it out to the country.

Cherry, Lynne. *A River Ran Wild: An Environmental History*. Boston, MA: HMH Books for Young Readers, 2002.

The author beautifully illustrates and tells the story of the restoration of the Nashua River in Massachusetts. The book follows the history of the river from the Native American settlement through the industrial revolution and pollution of the river, then final cleanup.

Cooney, Barbara. *Miss Rumphius*. New York: Puffin Books, 1985.

Miss Rumphius travels the world and then settles by the sea. There she plants lupine seeds to make the world a more beautiful place. The themes of movement, place, and human/environment interaction can be found in this beautifully illustrated book.

Gibbons, Gail. *How a House Is Built*. New York: Holiday House, 1990.

From the architect's plans through the arrival of a family, this book explains the process of construction of a house. Students will be introduced to carpenters, electricians, plumbers, and landscapers.

Golden, Nancy. *Exploring the United States with the Five Themes of Geography*. New York: The Rosen Publishing Group, Inc., 2005.

Where is the United States located? What is it like in the United States? How do the people and the environment of the United States affect one another? How do people, goods, and ideas get from place to place in the United States? What do places within the United States have in common to make them part of a region? These questions are thoroughly answered in this book through text and photographs.

Morris, Ann. *Houses and Homes* (Around the World Series). New York: HarperCollins, 1995.

Photographs of houses from all over the world show how the landscape features and resources of an environment were used to build the homes. Materials used include stone, wood, mud, and straw.

Say, Allen. *Grandfather's Journey*. Torrance, CA: Sandpiper Books, 2008.

This beautiful story tells how an immigrant from Japan travels through America by riverboat, train, and foot and sees various people, deserts, oceans, and rural and industrial towns. He misses his home in Japan, so he journeys back there, only to find out that he misses his home in America.

Seuss, Dr. *The Lorax*. New York: Random House Books for Young Readers, 1971.

Dr. Seuss promotes an ecological warning of the dangers of greed and human consequences on our environment. Trees are chopped down until only one is left. The book teaches that one small seed or one small child can make a difference in the world.

Yerrill, Gail. *The Gingerbread Man*. Bath, England: Parragon Books, 2018.

The gingerbread man runs from the old woman, the old man, the pig, the cow, and the horse, until he meets a sly fox. This classic story is modernized with large fonts and contemporary illustrations.

ASSESSMENT QUESTIONS—MAPPING WITH THE FIVE THEMES OF GEOGRAPHY

- Describe your hometown by location, place, and region.
  Possible Answer: My hometown is located at 38.0406° N, 84.5037° W. It is in the northeast part of the state and part of the Bluegrass Region.
- Name three ways humans interact with the environment.
  Possible Answer: Humans build roads and bridges, plant crops, and build dams.

- Name three ways the environment affects humans.
  Possible Answer: The environment affects the materials available to build homes, the type of clothing worn by the inhabitants, and the types of industry that can be located in an area.
- Choose two regions and describe how they are alike and how they are different.
  Possible Answer: The Northeast region of the U.S. has a continental climate with cool summers. The West region of the U.S. has a range of climates. California has a Mediterranean climate, while desert climates can be found in Nevada. The regions are similar because they are bordered by oceans. The Pacific Ocean borders the states on the West. The Northeastern states are bordered by the Atlantic Ocean.
- Besides goods, what else moves?
  Possible Answer: People move as they migrate or immigrate. Ideas move through communications. Water moves throughout the oceans and rivers.
- Why do you think people first settled in your hometown? Use the five themes of geography in your answer.
  Possible Answer: I think that people settled in this area because two rivers pass here. There is also a railroad through my town. The rivers and rails provided transportation means and water and food supply. The climate is temperate, and a coal field is nearby.
- How can England and the U.S. be part of the same region?
  Possible Answer: Both England and the U.S. are a part of the northern hemisphere.
- What is a good mnemonic for remembering the five themes of geography? Many Legal People Really Have Energy. Can you create another mnemonic?
  Possible Answer: MR. HELP is a mnemonic for remembering the five themes of geography. Legal People Really Have Much Energy is a mnemonic I created.
- Give examples of describing a place by physical characteristics, then by human/cultural characteristics.
  Possible Answer: The place where I live has mostly flat land that is below sea level. In this place, people love to eat spicy foods and celebrate Mardi Gras.

- Why do we divide places into regions?

  Possible Answer: Geographers divide places into regions based on information they want to study. Regions help geographers understand human and physical patterns. Some types of regions might include the Midwest, Uptown, the Alps, fruit-growing areas, or Cajun country.

- Give an example for each of the five themes of geography for your home area.

  Possible Answer: Movement—ships come into the port of New Orleans and goods travel across the ocean to other ports. Region—the French triangle between New Orleans, Lafayette, and Lake Charles houses many people of the Cajun culture. Human/Environment Interaction—there are many levees built around New Orleans because it is below sea level. The hurricanes destroyed many districts around the city, and some are now abandoned. Place—New Orleans has a lot of architecture and street names that represent both the Spanish and French culture. The food and music in the area have their own unique flavor and style. Location—New Orleans is located at 30 degrees N and 90 degrees W.

NOTES

1. Joint Committee on Geographic Education, *Guidelines for Geographic Education: Elementary and Secondary Schools* (Washington, D.C.: National Council for Geographic Education and Association of American Geographers, 1984).

2. Carol E. Murphey, "Using the Five Themes of Geography to Explore a School Site," *Journal of Geography* 90, no. 1 (1991): 38–40, DOI: 10.1080/00221349108979228.

3. Lesson idea from "Place," Smithsonian Institution Traveling Exhibition Service (SITES), *Earth 2U: Exploring Geography Curriculum Guide* (Washington, D.C.: Smithsonian Institute and the National Geographic Society, 1996), 9.

4. Lesson idea from "Place," *Earth 2U*, 9.

5. Lesson idea from "Human/Environment Interaction," *Earth 2U*, 10.

6. Lesson idea from "Movement," *Earth 2U*, 11.

7. Lesson idea from "Regions," *Earth 2U*, 12.

8. Lesson idea from William D. Edgington and William Hyman, "Using Baseball in Social Studies Instruction: Addressing the Five Fundamental Themes of Geography," *The Social Studies* 96, no. 3 (2005): 113–17.

9. Adapted from activity by Cheryl S. Knight, *Five Themes of Geography: Geography across the Curriculum: A Teacher's Activity Guide (Grades K–5)* (Boone, NC: Parkway Publishers, 1994), 228.

10. Lesson idea from Ava L. McCall, "Promoting Critical Thinking and Inquiry through Maps in Elementary Classrooms," *The Social Studies* 102, no. 3 (2011): 132–38.

11. Lesson idea from Victoria B. Fantozzi, Elizabeth Cottino, and Cindy Gennarelli, "Mapping Their Place: Preschoolers Explore Space, Place, and Literacy," *Social Studies and the Young Learner* 26, no. 1 (2013): 5–10.

12. Adapted from lesson by Pat Robeson, teacher consultant with the Maryland Geographic Alliance.

13. Lesson idea from "Five Themes," *Earth 2U*, 13.

14. Adapted from activity by Knight, *Five Themes of Geography*, 4.

15. Adapted from activity by Knight, *Five Themes of Geography*, 89.

Moving Forward with Geography

More than twenty-five years ago, the National Geographic Society secured the establishment of Geography Awareness Week,[1] the third week of November each year. The society was concerned about the deficiencies in American education of geography and its connections to everyday life. Community members, schools, and families are encouraged to participate in geographic activities during the week and celebrate their love of geography. Materials and resources for the celebration of Geography Awareness Week can be found at the National Geographic Society webpage.

The National Geographic Society also sponsors the annual academic competition for the Geography Bee (GeoBee).[2] Students in fourth through eighth grades can compete in the GeoBee. Teachers can register their schools to participate in the GeoBee through December of each year. The school champions must take an online test, and the top one hundred ranked students in each state qualify to represent their schools in the state championship. The state champions then attend the national championship in Washington, D.C.

Teaching geography will help create citizens who are concerned about major issues and problems facing the world: climate change, hunger, poverty, terrorism, degrading infrastructure, energy dependence, violent conflicts, environmental damage, and globalization. Many jobs today require workers to travel beyond their own states or nations. It is critical that we create students with the geographic knowledge and skills they need to comprehend the physical and cultural changes in the world. Geospatial technology is one of the fastest growing career areas available, as it is used in environmental fields, social services, real estate, government jobs, and the military.[3]

Introducing mapping skills allows teachers to help students develop spatial thinking skills. This allows students to visualize objects, find meaning in shape and orientation, and become aware of distance and relationships. Students can then answer the "where," the "how," and the "why there" questions that are a part of geographic thinking. Spatial thinking can then translate into other areas of life, such as how to arrange groceries in a bag, maneuver around a traffic jam, or give directions to a place.

With the ever increasing diversity in schools, the geography classroom can become the primary place where students can learn about cultures and countries, patterns of economic interdependence, and issues that will affect the livelihoods of all citizens. There is a lack of funding and lack of accountability for geography education. It is, therefore, up to teachers to ensure that students are able to ask geographic questions and construct geographic explanations. It is these young geographers who will go forth and face the major challenges and determine how to maintain the health of the world.

NOTES

1. Geography Awareness Week information can be found at https://www. nationalgeographic.org/education/programs/geography-awareness-week/.

2. GeoBee information can be found at https://www.nationalgeographic. org/education/student-experiences/geobee/.

3. *Why Geography Is Important* (San Marcos, TX: Gilbert M. Grosvenor Center for Geographic Education, 2012).

Appendix A
Suggested Scope and Sequence for Map Skills[1]

KINDERGARTEN

- Determine directions such as up, down, over, under, beside, forward, backward
- Compare distances such as long, short, near, far
- Recognize a globe as a model of the earth
- Use terms that express relative size such as big, little, large, small
- Recognize that models and symbols represent real things
- Identify the local community

GRADE 1

- Differentiate between maps and globes
- Interpret simple map symbols using a legend
- Recognize land and water representations on maps and globes
- Create simple maps of the local environment
- Use simple maps to locate objects
- Know location of home in relation to school

GRADE 2

- Identify physical shapes of country and states
- Identify cardinal directions
- Compare pictures and maps of the same area
- Identify local landforms

- Make maps of the school and neighborhood
- Use geographical terms to describe what you see (e.g., hill, slope, river)

GRADE 3

- Use a map or globe to locate continents and oceans
- Locate points on a map or globe using directions (NE, NW, SE, SW)
- Use scale to determine distance on a simple map
- Compare own community with other communities
- Compare rural and urban environments
- Make representations of imaginary places

GRADE 4

- Locate the prime meridian, equator, and international date line on a map and globe
- Use special purpose maps to identify and gather data (e.g., climate, product, population)
- Identify major physical features of the world
- Work with distance, direction, scale, and map symbols
- Recognize common characteristics of map grids and map projections
- Follow a route map created by the teacher

GRADE 5

- Identify states and capitals of the United States
- Identify the regions of the United States and major physical features
- Identify neighboring countries on a map
- Collect data from maps using longitude, latitude, legends, and scales
- Discuss location in terms of where and why
- Use the eight points of a compass
- Use letter and number coordinates to locate features on a map

GRADE 6

- Identify a geographic region within a nation
- Identify geographic data to draw conclusions, defend conclusions, and make predictions
- Distinguish between a political and physical map
- Map trade routes
- Plot distributions of population and key resources on maps
- Examine how people have modified and adapted to the environment

NOTE

1. Adapted from "The Role and Sequence of Geographic Education," *Guidelines for Geographic Education: Elementary and Secondary Schools*, prepared by the Joint Committee on Geographic Education of the National Council for Geographic Education and the Association of American Geographers (1984): 11–17.

Appendix B
Glossary of Geographic Terms

absolute location. The exact location noted in terms of latitude and longitude.

aerial photograph. A picture taken from above, typically from an airplane.

cartographer. A person who draws maps.

climate. Average weather over a period of time.

compass. An instrument used to find directions; the needle points to magnetic north.

compass rose. Figure on a map or compass designed to show directional terms.

contour line. A line on a topographic map that joins places of the same elevation or depth.

coordinate. The representation of the degree and direction of a latitude or longitude line.

degree. Measurement for latitude and longitude.

equator. An imaginary line at 0° that circles the earth and divides it into the northern and southern hemispheres.

geography. The study of location and place on the earth and natural and human activity.

globe. A representation of the earth.

grid. A series of equally spaced vertical and horizontal lines.

hemisphere. Half of the globe; the equator divides the earth into northern and southern hemispheres, while the prime meridian divides the earth into eastern and western hemispheres.

human/environment interaction. Used as a theme of geography to explain how humans affect the environment and the environment affects humans.

international date line. An imaginary line at 180° that indicates the place where the date changes.

key. A legend that explains the symbols on a map.

latitude. The distance north and south of the equator; imaginary lines sometimes called parallels.

legend. The key for decoding the symbols on a map.

location. Used as a theme of geography to explain the position of a place on the earth.

longitude. Distance east or west of the prime meridian; imaginary lines sometimes called meridians.

magnetic north. Position near the true North Pole of the earth that slowly changes over the years.

map projection. A way of showing the round earth as a flat map.

meridian. An imaginary line circling the earth north and south to measure longitude.

movement. Used as a theme of geography to explain how people, ideas, and products get from place to place.

parallel. An imaginary line circling the earth east and west to measure latitude.

physical map. A map that shows the physical features of the earth.

place. Used as a theme of geography to explain natural and cultural features of an area.

political map. A map that shows location of major cities and boundaries of countries, states, and counties.

precipitation. A weather condition where something is falling from the sky (e.g., rain, snow, sleet, hail).

prime meridian. The 0° line of longitude that passes through Greenwich in England.

region. Used as a theme of geography to connect different places by natural and/or cultural features.

relative location. Noting where a place is located in relation to another place.

satellite. A spacecraft orbiting the earth that can provide pictures.

scale. The units on a map that show the units on the ground.

symbol. A shape or picture on a map that stands for a real feature.

temperature. The degree of warmth or coldness measured on a defined scale.

time zone. An international system of twenty-four areas that differ by one hour.

topographic map. A map that uses elevation contour lines and both natural and man-made features.

weather. Day-to-day conditions measured by temperature and precipitation.

Appendix C
Map Projection Comparison Chart

Questions	Map Name	Map Name	Map Name	Map Name	Map Name
Why use this projection?					
What distortions do you see?					
Who created the map, when, where, and why?					
Compare the size of Greenland, South America, and Africa on each projection map.					

Map Projection Chart

Appendix D
News Shapes the World

Instructions: Find an article and label the content type. Read and summarize. Cut the article into the shape of a continent. Glue the continents on construction paper to shape the world. Note that the article does not have to take place in the continent you chose.

Article Content (environmental news, climate/weather, politics, local news, state news, sports news, international news)	Continent Shape of News Article	Summary of Article

News Shapes the World

Appendix E
Evaluating Maps in Children's Literature Chart

Book Title	Is the place real or imaginary?	How accurate is the map?	How useful is the map?	What is missing from the map?	Does the map have a scale?

Evaluating Maps in Children's Literature Chart

Appendix F
Grid Paper

Appendix G
Bingo Card

B	I	N	G	O
⭐				
		⭐		
				⭐

BINGO Card

Appendix H
Topographic Map

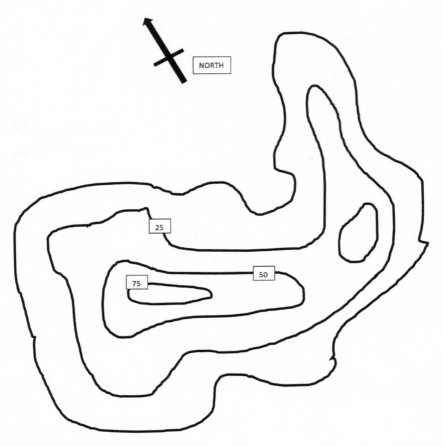

Topographic Map

Appendix I
Scavenger Hunt from Space Chart

Group Members	Lat/Long of Building	Clue #1	Clue #2	Clue #3

Scavenger Hunt from Space Chart

Bibliography

Alibrandi, Marsha. "How to Do It: Online Interactive Mapping: Basic Activities." *Social Studies and the Young Learner* 17, no. 3 (2005): P1–P8.

Bednarz, Sarah Witham, Gillian Acheson, and Robert S. Bednarz. "Maps and Map Learning in Social Studies." *Social Education* 70, no. 7 (2006): 398–404, 432.

Dykes, Jason, and Jo Wood. "Rethinking Map Legends with Visualization." *IEEE Transactions on Visualization and Computer Graphics* 16, no. 6 (2010): 890–99.

Edgington, William D., and William Hyman. "Using Baseball in Social Studies Instruction: Addressing the Five Fundamental Themes of Geography." *The Social Studies* 96, no. 3 (2005): 113–17.

"Esri Pledges $1B in Cloud-Based STEM Software to White House ConnectED Initiative." *The Esri Newsroom*. May 27, 2014. Accessed November 4, 2019, http://www.esri.com/esri-news/releases/14-2qtr/esri-pledges-1b-in-cloudbased-stem-software-to-white-house-connected-initiative.

Fantozzi, Victoria B., Elizabeth Cottino, and Cindy Gennarelli. "Mapping Their Place: Preschoolers Explore Space, Place, and Literacy." *Social Studies and the Young Learner* 26, no. 1 (2013): 5–10.

Fraser, Celeste. *Unlocking the Five Themes of Geography*. Englewood Cliffs, NJ: Globe Book Company, 1993.

Frazee, Bruce M. "Foundations for an Elementary Map Skills Program." *The Social Studies* 75, no. 2 (1984): 79–82.

Gandy, S. Kay. "Developmentally Appropriate Geography." *Social Studies and the Young Learner* 20, no. 2 (2007): 30–32.

Gandy, S. Kay. "Mapping Skills and Activities with Children's Literature." *Journal of Geography* 105, no. 6 (2006): 267–71, DOI: 10.1080/00221340608978696.

Getskow, Veronica. *Incredible Edible Geography.* Irvine, CA: Thomas Bros. Maps Educational Foundation, 1998.

Gregg, Madeleine. "Seven Journeys to Map Symbols: Multiple Intelligences Applied to Map Learning." *Journal of Geography* 96, no. 3 (1997): 146–52.

Haas, Mary. "Teaching Geography in the Elementary School." *ERIC Digest* (1989): 2–7, ED309133.

Hannibal, Mary Anne Zeitler, Ren Vasiliev, and Qiuyun Lin. "Teaching Young Children Basic Concepts of Geography: A Literature-Based Approach." *Early Childhood Education Journal* 30, no. 2 (2002): 81–86.

Haslam, Andrew. *Maps: Make It Work!* Chicago: World Book Inc., 1996.

"History of Mapping." Intergovernmental Committee on Surveying and Mapping (ICSM). Accessed November 4, 2019, https://www.icsm.gov.au/education/fundamentals-mapping/history-mapping.

Joint Committee on Geographic Education. *Guidelines for Geographic Education: Elementary and Secondary Schools.* Washington, D.C.: National Council for Geographic Education and Association of American Geographers, 1984.

Kelly, Kate. *That's Not in My Geography Book: A Compilation of Little-Known Facts.* New York: Taylor Trade Publishing, 2009.

Kenda, Margaret, and Phyllis S. Williams. *Geography Wizardry for Kids.* New York: Scholastic, Inc., 1997.

Knight, Cheryl S. *Five Themes of Geography: Geography across the Curriculum: A Teacher's Activity Guide (Grades K–5).* Boone, NC: Parkway Publishers, 1994.

"Making an Earth Sandwich." *Explore Your World with a Geographic Information System: A Teaching Supplement for Grades 5–12 Introducing Basic GIS Concepts and Components.* Environmental Systems Research Institute, Inc., 1995. Accessed November 4, 2019, https://www.scribd.com/document/22344623/explore-your-world-with-a-GIS-teachers.

Malone, Lyn, Anita M. Palmer, and Christine L. Voigt. *Mapping Our World: GIS Lessons for Educators.* Redlands, CA: ESRI Press, 2002.

"Mapping a London Epidemic." National Geographic Education. Accessed November 4, 2019, https://www.nationalgeographic.org/activity/mapping-london-epidemic/.

Mason, Charlotte. *Elementary Geography.* London: Edwards Stanton, 2015.

McCall, Ava L. "Promoting Critical Thinking and Inquiry through Maps in Elementary Classrooms." *The Social Studies* 102, no. 3 (2011): 132–38.

Muir, Sharon Pray. "Understanding and Improving Students' Map Reading Skills." *Elementary School Journal* 86, no. 2 (1985): 207–16.

Murphey, Carol E. "Using the Five Themes of Geography to Explore a School Site." *Journal of Geography* 90, no. 1 (1991): 38–40, DOI: 10.1080/00221349108979228.

Palmer, Joy. *Geography in the Early Years.* New York: Routledge, 1994.

Parnell, Elysia. "Geography Is All about Location, Location, Location!" *Teaching Geography* 32, no. 2 (2007): 91–92.

Rhatigan, Joe, and Heather Smith. *Geography Crafts for Kids: 50 Cool Projects & Activities for Exploring the World.* New York: Scholastic, Inc., 2002.

Richards, Betty. "Mapping: An Introduction to Symbols." *Young Children* 31, no. 2 (1976): 145–56.

Schoenfeldt, Melinda. "Geographic Literacy and Young Learners." *The Educational Forum* 66, no. 1 (2001): 26–31, DOI: 10.1080/00131720108984796.

Segall, Avner. "Maps as Stories about the World." *Social Studies and the Young Learner* 16, no. 1 (2003): 21–25.

Shin, Eui-kyung. "Using Geographic Information System (GIS) Technology to Enhance Elementary Students' Geographic Understanding." *Theory & Research in Social Education* 35, no. 2 (2007): 231–55, DOI: 10.1080/00933104.2007.10473334.

"Shipwreck Island." *Geography Assessment Framework for the 1994 National Assessment of Educational Progress.* NAEP Geography Consensus Project. Contract Number RN91073001. Council of Chief State School Officers, National Council for Geographic Education, National Council for the Social Studies, American Institutes for Research, 1994.

Smithsonian Institution Traveling Exhibition Service. *Earth 2U: Exploring Geography Curriculum Guide.* Washington, D.C.: Smithsonian Institution and the National Geographic Society, 1996.

Sobel, David. *Mapmaking with Children: Sense of Place Education for the Elementary Years.* Portsmouth, NH: Heinemann, 1998.

Swartz, David J. "A Study of Variation in Map Symbols." *The Elementary School Journal* 33, no. 9 (1933): 678–79.

"The Role and Sequence of Geographic Education." *Guidelines for Geographic Education: Elementary and Secondary Schools.* Joint Committee on Geographic Education of the National Council for Geographic Education and the Association of American Geographers, 1984.

VanCleave, Janice. *Geography for Every Kid: Easy Activities That Make Learning Geography Fun.* New York: John Wiley & Sons, Inc., 1993.

"What Is the Prime Meridian and Why Is It in Greenwich?" Royal Museums Greenwich. Accessed November 1, 2019, http://www.rmg.co.uk/discover/explore/prime-meridian-greenwich.

"Where There's Light, There Are People." *Teacher's Handbook for Geography Awareness Week*. Washington, D.C.: National Geographic Society, 1998.

"World Population Map: A Cartogram for the Classroom." Population Education. Accessed November 4, 2019, https://populationeducation.org/world-population-map-cartogram-classroom/.

Summary List of Activities

Summary List of Activities

CHAPTER 1: Location

#	Activity	Time Needed	Page	Suggested Grade Levels						
				K	1	2	3	4	5	6
1	Toss the Globe	15 minutes	3	X	X	X	X	X	X	X
2	Your Place in the World	30 minutes	4		X	X				
3	Where Are You Now?	45 minutes	5		X	X	X	X	X	
4	Where Are You Going? Where Have You Been?	25 minutes	6		X	X	X	X	X	
5	Travel Stories	45 minutes	6					X	X	X
6	Music across America	20 minutes	7					X	X	X
7	Where Do Products Come From?	25 minutes	8				X	X	X	X
8	Family News Night	20 minutes	9					X	X	X
9	Route Map	30 minutes	10				X	X	X	X
10	MIMAL—The Man in the Middle	15 minutes	10				X	X	X	
11	We Are the World	40 minutes	11			X	X	X		

CHAPTER 2: Perspective						Suggested Grade Levels					
#	Activity	Time Needed	Page	K	1	2	3	4	5	6
12	Earth Models	45 minutes	19			X	X	X		
13	Flat Map vs. Globe	45 minutes	19				X	X	X	X
14	Food as Earth	45 minutes	20			X	X	X		
15	Map Projections	30 minutes	20						X	X
16	View from the Top	45 minutes	21		X	X				
17	Earth Shapes	45 minutes	22					X	X	X
18	Learning through Play	30 minutes	23	X	X	X				
19	Mental Mapping	30 minutes	23					X	X	X
20	How Big Is Africa?	30 minutes	24					X	X	X
21	News Shapes the World	45 minutes	24					X	X	X
22	Maps and Math	20 minutes	25					X	X	X
23	Body Mapping	25 minutes	25			X	X	X		

CHAPTER 3: Scale				Suggested Grade Levels						
#	Activity	Time Needed	Page	K	1	2	3	4	5	6
24	A Scale of Two Cities	25 minutes	30					X	X	X
25	Mapping the Classroom	30 minutes	31				X	X	X	
26	Evaluating Maps	30 minutes	31					X	X	X
27	Literature and Maps	45 minutes	32				X	X	X	X
28	Tabletop Maps	30 minutes	32			X	X	X		
29	Big to Small	30 minutes	33		X	X	X			
30	Blueprints	45 minutes	33					X	X	X
31	Shapes to Scales	20 minutes	34				X	X	X	
32	Scale Drawing	45 minutes	34					X	X	X
33	Move the Room	45 minutes	35					X	X	X
34	Jigsaw Puzzle	45 minutes	35					X	X	X

CHAPTER 4: Orientation

#	Activity	Time Needed	Page	Suggested Grade Levels						
				K	1	2	3	4	5	6
35	Cardinal Directions	30 minutes	41	X	X	X				
36	Orientation Games	20 minutes	41	X	X	X				
37	Pirate Treasure Hunt	45 minutes	42				X	X	X	
38	Reading a Compass	30 minutes	43			X	X	X	X	X
39	Hemispheres	15 minutes	43	X	X	X				
40	Time Zones	30 minutes	44					X	X	X
41	What Time Is It?	30 minutes	44					X	X	X
42	Upside-Down World	20 minutes	45			X	X	X		
43	Which Way Is North?	45 minutes	45					X	X	X
44	Building a Compass Rose	30 minutes	46	X	X	X				
45	Use Your Brain to Navigate	1 day	46						X	X

CHAPTER 5: Map Symbols and Map Keys				Suggested Grade Levels						
#	Activity	Time Needed	Page	K	1	2	3	4	5	6
46	What Is It?	30 minutes	53	X	X					
47	Map Symbol Bingo	30 minutes	53			X	X	X		
48	Creating Symbols	30 minutes	54		X	X	X			
49	A Symbol by Any Other Name	45 minutes	54					X	X	X
50	Neighborhood Symbols	1 hour	55			X	X	X		
51	Matching Symbols	30 minutes	56	X	X					
52	Points, Lines, and Areas	25 minutes	56			X	X	X		
53	Game On!	30 minutes	56					X	X	X
54	A Legend in Its Own Time	30 minutes	57					X	X	X
55	Cookie Symbols	30 minutes	57		X	X				
56	Mapping Mnemonics	25 minutes	58				X	X	X	X

CHAPTER 7: Mapping with Technology

#	Activity	Time Needed	Page	Suggested Grade Levels						
				K	1	2	3	4	5	6
73	Geocaching	1 to 3 hours	82						X	X
74	EarthCaching	1 to 3 hours	83						X	X
75	GIS Sandwich	45 minutes	83				X	X	X	X
76	GeoScavenger Hunt	1 day or weekend	84				X	X	X	X
77	Earth from Space	30 minutes	84				X	X	X	X
78	Batty Images	30 minutes	85			X	X	X		
79	Story Mapping	Several days	85					X	X	X
80	My Home, School, and Community	1 hour	86					X	X	X
81	MapQuest	45 minutes	86				X	X	X	X
82	The Amazing Race	2 hours	87					X	X	X
83	Scavenger Hunt from Space	1 hour	87				X	X	X	X
84	Community Life	1 hour	88					X	X	X
85	Learning How to Pan	45 minutes	88		X	X	X			
86	Learning How to Zoom	25 minutes	89		X	X	X			

#	Activity	Time Needed	Page	Suggested Grade Levels						
				K	1	2	3	4	5	6
87	Where to Build?	20 minutes	95					X	X	X
88	Guessing Locations	20 minutes	96					X	X	X
89	Positive/Negative	25 minutes	96					X	X	X
90	Space Station on the Moon	30 minutes	97					X	X	
91	What Is a Region?	30 minutes	97			X	X	X		
92	Build Your House	25 minutes	98			X	X	X		
93	Baseball and the Five Themes	45 minutes	98						X	X
94	Five Themes Book Map	30 minutes	99				X	X	X	X
95	My Special Place	30 minutes	99			X	X	X		
96	Playground Project	1 afternoon	100					X	X	X
97	3D Playground	1 afternoon	100	X	X	X				
98	Five Themes Photographs	30 minutes	101				X	X	X	
99	The Three Bears	30 minutes	102	X	X					
100	Find the Gingerbread Man	1 hour	102			X	X	X		

CHAPTER 8: Mapping with the Five Themes of Geography

About the Author

Before her work in higher education, **Dr. S. Kay Gandy** taught in public schools in Louisiana for twenty-seven years. She has worked to train teacher candidates for the past seventeen years in Kentucky. Dr. Gandy is a perpetual student and has five degrees, including a master of science in geography. Her passion is international education, and she has created many opportunities for both faculty and students to travel abroad. She has received two Fulbright awards to South Africa and one to Senegal, and she is currently serving as a Fulbright Specialist. In the 1990s, Dr. Gandy trained with the National Geographic Society to serve as a teacher consultant with the Louisiana Geography Education Alliance (LaGEA), and later served as a co-coordinator for the Kentucky Geographic Alliance.